Organic Dye-Doped
PMMA
Thin-Film Laser

Organic Dye-Doped PMMA Thin-Film Laser

Von der Fakultät für Elektrotechnik, Informationstechnik, Physik

der Technischen Universität Carolo-Wilhelmina zu Braunschweig

zur Erlangung des Grades eines Doktors

der Ingenieurwissenschaften (Dr.-Ing)

genehmigte Dissertation

von Ang Pen Yiao

aus Pulau Pinang, Malaysia

eingereicht am: 25.06.2021

mündliche Prüfung am: 01.09.2021

1. Referent: Prof. Dr. Wolfgang Kowalsky
2. Referent: Prof. Dr. Bernhard Wilhelm Roth

Druckjahr: 2021

Bibliografische Information der Deutschen Nationalbibliothek
Die Deutsche Nationalbibliothek verzeichnet diese Publikation in der Deutschen Nationalbibliografie; detaillierte bibliographische Daten sind im Internet über http://dnb.d-nb.de abrufbar.
1. Aufl. - Göttingen: Cuvillier, 2021
Zugl.: (TU) Braunschweig, Univ., Diss., 2021

Dissertation an der Technischen Universität Braunschweig
Fakultät für Elektrotechnik, Informationstechnik, Physik

Nonnenstieg 8, 37075 Göttingen
Telefon: 0551-54724-0
Telefax: 0551-54724-21
www.cuvillier.de

1. Auflage, 2021
Gedruckt auf umweltfreundlichem, säurefreiem Papier aus nachhaltiger Forstwirtschaft.

ISBN 978-3-7369-7503-3
eISBN 978-3-7369-6503-4

List of Publication

With the permission of the Faculty of Electrical Engineering, Information Technology, Physics of Technical University Braunschweig represented by the mentor of the work, parts of this dissertation have been published in peer-reviewed scientific journals and on international conferences.

Journal

- P. Y. Ang, M. Čehovski, F. Lompa, C. Hänisch, D. Samigullina, W.Kowalsky, H.-H. Johannes. "Organic dye-doped PMMA lasing." *Polymers* 13.20 (2021): 3566.
- H.-A. Christ, P. Y. Ang, F. Li, W. Kowalsky, H.-H. Johannes, H. Menzel. "Production of highly aligned microfiber bundles from PMMA via stable jet electrospinning for organic solid-state lasers." *In proceedings.*
- S. Spelthann, S. Unland, J. Thiem, F. Jakobs, J., Kielhorn, P. Y. Ang, H.-H. Johannes, D. Kracht, J. Neumann, A. Ruehl, W. Kowalsky, D. Ristau. "Towards Highly Efficient Polymer Fiber Laser Sources for Integrated Photonic Sensors." *Sensors* 20.15 (2020): 4086.
- F. Jakobs, K. Harms, J. Kielhorn, D. Zaremba, P. Y. Ang, W. Kowalsky, H. H. Johannes. "Homogeneous Distribution of Polymerizable Coumarin Dyes for Active Few Mode POF." *Materials* 13.8 (2020): 1975.

Book

- P. Y. Ang, J. Kielhorn, F. Jakobs, W. Kowalsky, H.-H. Johannes. "Lumogen-doped polymer fibers", *Drittes Deutsches POF-Symposium 2020*, Tewiss Verlag, 2020, ISBN: 978-3-95900-442-8

Conferences

- P. Y. Ang, J. Kielhorn, F. Jakobs, R. Caspary, W. Kowalsky and H.-H. Johannes. "Lumogen Dye-Doped Polymer Optical Fibers", *28th International Conference on Plastic Optical Fibers (POF2019)*, Yokohama (Japan), NOV. 20-22, 2019.

- J. Kielhorn, P. Y. Ang, F. Jakobs, W. Kowalsky and H.-H. Johannes. "New Organic Dyes integrated into Polymer Optical Fibers", *28th International Conference on Plastic Optical Fibers (POF2019)*, Yokohama (Japan), NOV. 20-22, 2019.

- R. Caspary, F. Jakobs, J. Kielhorn, P. Y. Ang, M. Cehovski, M. Beck, H.-H. Johannes, S. Balendat, J. Neumann, S. Unland, S. Spelthann, J. Thiem, A. Ruehl, D. Ristau, W. Kowalsky. "Polymer fiber laser", *21th International Conference on Transparent Optical Networks (ICTON)*, Angers (France), JUL. 9-13, 2019.

Abstract

Organic dye laser stands a chance to become the next generation of the light source. It is very interesting because organic dyes can cover a wide range of the spectrum of visible light. To realise organic solid-state laser, several organic dyes can be doped in a polymer matrix. One of the examples of low-priced polymer is poly(methyl methacrylate) (PMMA). However, the study of PMMA as a polymer matrix for organic dyes as a thin-film laser was not extensive. This thesis focused on the feasibility of organic dye-doped PMMA as a thin-film laser.

Six different sample structures were suggested to determine the best sample model with the best lasing properties. First, the structure consisted of a merely dye-doped thin-film. In the second design, a grating structure was built directly on top of the gain medium. Third, the dye-doped thin-film was covered with a layer of photoresist, EpoClad. In the fourth design, the grating structure was established on top of the gain medium and enclosed by the EpoClad layer at the same time. Next, the dye-doped thin-film was sandwiched between the substrate and the EpoClad layer in the fifth design, in which the grating structure was produced on top of the EpoClad layer. In the sixth design, the dye-doped PMMA was spin-coated on top of an EpoClad layer with a grating structure. Lasing properties like narrowing of Full Width at Half Maximum (FWHM) and a clear lasing threshold were focused. In this experiment, Rhodamine 6G was selected as the dye-dopant for PMMA thin-film because it was commonly used for fiber lasers. It was discovered that the fifth design presented the best performance with the smallest FWHM, the highest polarisation extinction ratio (PER) and the highest output energy. To ensure good lasing performance, the thickness of the EpoClad layer and the Rhodamine 6G doped PMMA layer was manipulated. The best sample structure had an EpoClad layer thickness of 661 nm and a PMMA layer thickness of 1200 nm. With the sample model, further experiments proceeded.

On top of the detection of a small FWHM and a clear lasing threshold, changes of the gain medium and the resonator should be proceeded to assure the laser state. At first, the Rhodamine 6G concentration in PMMA was varied to check its influence. When the dye concentration was too low (100 ppm), no lasing properties were determined. The best result was measured by the sample doped with 400 ppm Rhodamine 6G concentration. Therefore, the dye concentration was fixed at 400 ppm for further experiments. Second, changes in the grating period of the grating structure on the sample would influence the lasing properties because the grating structure functioned as the resonator. Lasing properties were observed when the grating period was in the range from 370 nm to 390 nm. With the variation of grating periods, a 25 nm of wavelength shift was measured.

Moreover, the workability of PMMA with other organic dyes was also investigated. It was noticed that there were in total six organic dyes, which could function well with PMMA as an organic dye laser. They were Rhodamine 6G (Rh6G), Rhodamine B (RhB), Lumogen Orange (LumO), Pyrromethene 597 (P597), 4-(Dicyanomethylene) -2-tert-butyl-6- (1,1,7,7-tetramethyljulolidin-4-yl-vinyl)- 4H-pyran (DCJTB) and 4-(Dicyanomethylene)-2-methyl-6-julolidyl-9-enyl-4H- pyran (DCM2). By using these dyes, the laser peak wavelength was found at the range from 572 nm to 609 nm. Among these dyes, Pyrromethene 597 doped sample showed the best result with a peak wavelength of 572.45 nm, a FWHM of 0.57 nm, a lasing threshold of 0.253 μJ, a PER of >30 dB, and an output energy of 18 nJ. The P597 doped sample was modified to detect sugar solution. When the sugar concentration increased from 0wt% to 40wt%, a 4 nm of wavelength shift was detected.

It can be concluded that an organic dye-doped PMMA thin-film can work as an organic dye laser. The best sample design was found, in which the dye-doped PMMA thin-film was sandwiched in between the substrate and the EpoClad layer, where the grating structure was built on top of the EpoClad layer. A narrow FWHM (<1 nm) and an obvious lasing threshold were observed. The lasing properties were affected by the modification of the gain medium and the resonator. Six various organic dyes could be doped in PMMA to function as an organic dye-doped PMMA thin-film laser. The sample stood the potential to detect e.g. sugar concentration. Further experiments should be carried on for more applications.

Contents

1 Introduction

The role of organic materials is essential in our daily life. In former times, the only possibility to achieve light was to ignite woods or oil, which are organic materials. Nowadays, organic semiconductors have gained extensive attraction due to their large-area and low-cost fabrication. Besides, the optoelectronic properties of organic semiconductors can be easily tuned by their convenient molecular design, which increases their potential to be applied in various kinds of electronics products.

Organic semiconductors are been an important material for numerous devices such as organic light-emitting diodes (OLEDs) [1–3], organic solar cells (OSCs) [4, 5], organic thin-film transistors [6, 7] and organic solid-state lasers (OSSLs) [8, 9]. Since there is high interest in the industry and technology field, a variety of organic semiconductors have been researched. Especially, OLEDs are been strongly promoted into commercial applications as the display of smartphones and televisions. OSSLs have the possibility to be another organic light emitter. Yet, it is still a young and challenging research topic. OSSLs provide another new horizon for simple, low-cost, time-saving, versatile, and environmentally-friendly fabrication of new and desirable laser structures. There is numerous literature about OSSLs. Diverse application fields have been found, such as lab-on-chip spectroscopy [10], absorption and transmission spectroscopy [11, 12], data/optical communication [13], refractive index sensor [14], vapour pressure detector [15] and etc.

The first-ever OSSL was discovered in 1967 when Rh6G doped in poly(methyl methacrylate) (PMMA) was found to be an organic gain medium [16]. Afterwards, much effort was done into this topic. One of the organic semiconductor lasers (OSLs) was reported, in which the materials were anthracene in doped and non-doped single crystals [17]. The first thin-film OLED was proposed by Tang and Van Slyke in 1987 [18]. This significant achievement of OLEDs contributes to a better understanding of the structural design and functions of organic materials. Subsequently, this leads to the significant development of OSLs. In 1996, Friend and Heeger introduced an OSL based on conjugated polymers by using poly(p-phenylenevinylene) (PPV) and PPV derivatives as gain medium [19, 20]. Later, Samuel's group introduced a polymer laser pumped by LED, in which a fluorene copolymer was used as the gain media [21]. In 2009, a deep-blue laser with a very low lasing threshold and high slope efficiency using monodisperse starburst macromolecular semiconductor as emitter was proposed [22]. A few years later, Gather et al. discovered the first observation of lasing behaviour in a living system with green fluorescent protein [23]. Recently, Adachi et al. demonstrated a quasi continuous-wave OSL based on near-infrared thermally activated delayed fluorescent emitters and the possibility of an electrically pumped organic laser [24–26]. Figure 1.1 shows the development milestones of organic gain media.

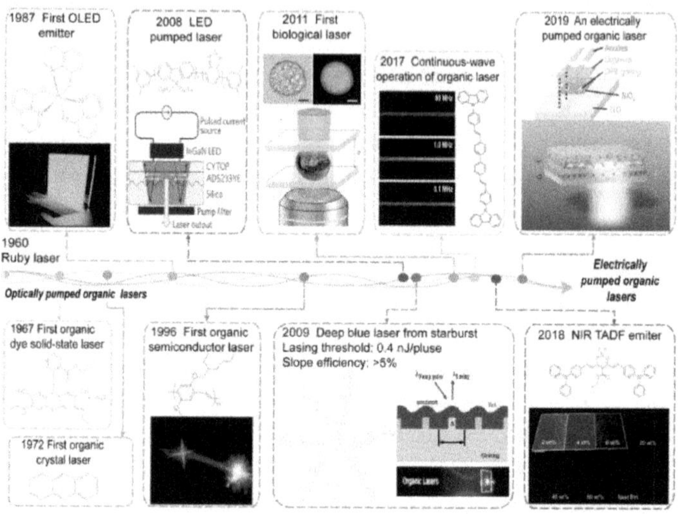

Figure 1.1: Timeline of development of organic main media.

To function as organic gain media, organic dyes have been dissolved in liquid solvents for the lasing purpose for decades [27]. However, this kind of operation can induce unnecessary evaporation of the solvents. To solve this problem, organic dyes can be doped in non-conjugated polymers, such as polystyrene (PS) and PMMA, which are suitable for OSSLs. This resulted laser is defined as organic dye lasers, which is quite similar to OSLs. According to Samuel el al., there are properties to make a division for organic semiconductors: (i) easy thin-film fabrication, (ii) high photoluminescence quantum yield (PLQY), and (iii) potential of charge transport [28]. There is abundant literature about organic dye lasers. It is observed that there are several sample designs for the dye-doped PS samples [14, 29–32]. On the other hand, the dye-doped PMMA samples were mostly based on only thin-film structure [33, 34].

In this work, the feasibility of organic dye-doped PMMA laser was studied. There were in total six different sample designs proposed to find out the sample model with the best performance. Next, the layer thickness of the gain medium and the resonator of the best sample model was manipulated to assure the outcome. Lasing properties like narrowing of Full Width at Half Maximum (FWHM) and lasing threshold were monitored. Besides, alternations of gain medium and resonator were performed to observe the effect on the lasing properties, in which the concentration of dye and the grating period were manipulated. Moreover, the workability of different organic dyes with PMMA was also determined. It was shown that organic dye-doped PMMA could function as a laser, which had an application potential.

2 General theory

In this chapter, one will get an overview of the general basics. At first, a short description of light propagation is elucidated here. Then, it is followed by a brief explanation of optical waveguides, which are an example of the medium for light propagation. Besides, general information about organic semiconductors is shown, as the material used in this research is mainly made up of organic material. Moreover, two important characteristics of organic materials are briefly described, which are the photoluminescence quantum yield and the fluorescence lifetime. Furthermore, a compact description of quenching is presented to understand the processes, which can reduce the intensity of fluorescence.

Every single section provides elementary theories, which provide adequate physics information for the readers to understand this work better. Detailed explanations can be found on the indicated references.

2.1 Light Propagation

Electromagnetic radiation can be defined as a form of energy. In this form of energy, all the waves of the electromagnetic field such as gamma rays, X-rays, ultraviolet radiation, visible light, infrared, radio waves, etc are included. According to Figure 2.1, visible light is only a very small portion of the total electromagnetic radiation. Based on the DIN 5031-7, visible light starts from 380 nm to 780 nm [35]. Radiation energy can be induced by the propagation of electromagnetic radiation, which can be calculated using the equation below:

$$E_{\text{radia}} = h\upsilon = \frac{hc}{\lambda} = \frac{hc}{2\pi}k \tag{2.1}$$

whereas E_{radia} is the radiation energy, h is the Planck constant, υ is the frequency, c is the velocity of light in a vacuum, λ is the wavelength and k is the wavenumber.

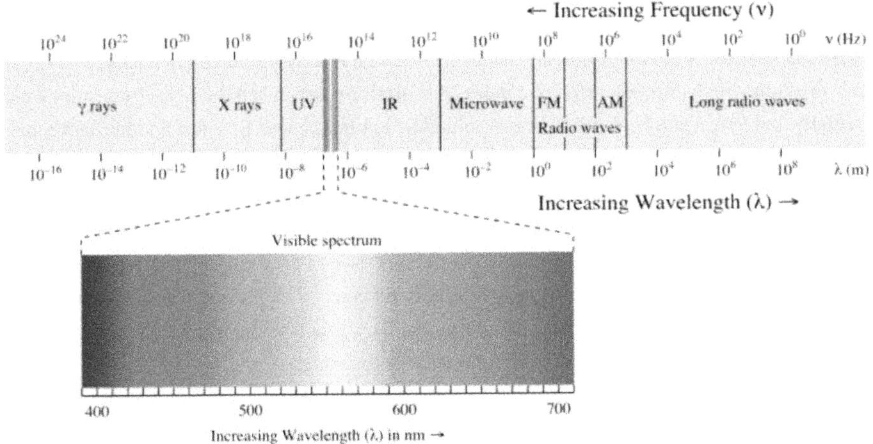

Figure 2.1: Electromagnetic spectrum with labelled regions. Reprinted from [36].

The concept of light propagation is proposed that the movement of the energy of the electromagnetic waves can be achieved from one point to another. Electromagnetic waves contain electric and magnetic components. Meanwhile, the light propagation is not the same as mechanical waves because no material is required for their energy transportation. Therefore, light is capable to pass through the vacuum of outer space with a velocity $c = 3 \times 10^8\,\mathrm{m\,s^{-1}}$ [37, 38]. The wavelength of the frequency of the light is in a relationship with the light velocity, which is shown in the following equation:

$$\upsilon = \frac{c}{\lambda}. \tag{2.2}$$

When there is light propagation in a medium, the velocity of light becomes less than the velocity in a vacuum. The actual velocity of light relies on the optical density of the material. This is because optical density is determined as the degree to which a medium can refract the transmitted light. In short, the whole mechanism is illustrated by the refractive index n_r. When the refractive index rises, the velocity of light in the medium decreases. n_r of vacuum is determined as 1, which is used as a reference. For instance, n_r of water is 1.33, which means the light propagation in water is 1.33 times slower compared to vacuum. n_r can be calculated using

$$n_r = \frac{c}{\nu} \tag{2.3}$$

whereas v is the phase velocity of light in the material.

Figure 2.2 presents an overview of the mechanisms when there is an interaction of light with another medium with a different refractive index. Absorption exists when the light is taken in, which is usually converted into heat if it is not reemitted as luminescence. Besides, another mechanism is called transmission when the electromagnetic waves are able to pass through the medium. After transmission, the angle of the light can be different compared to incident light as it is refracted. The angle of refraction is based on the Snell's law ($n_1 \sin(\theta_i) = n_2 \sin(\theta_{rr})$). On top of that, reflection takes place, when the light rebounds at the interface between two media with different refractive indices. Based on the law of reflection, the light is reflected with the same angle to the surface normal as the incident ray under the assumption that the surface of the interface is mirror-like ($\theta_i = \theta_{rl}$), otherwise, scattering may occur if the surface is too rough.

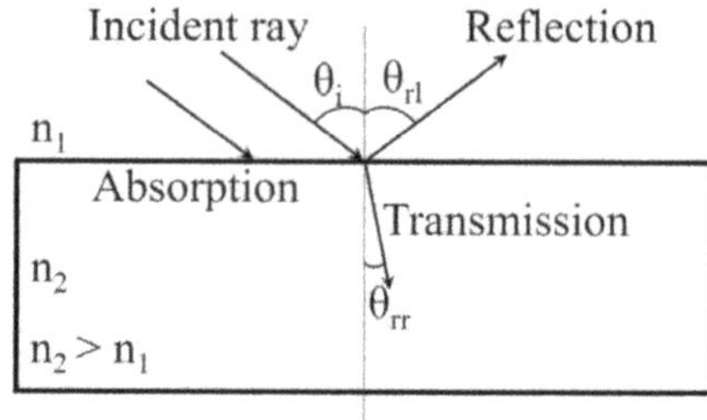

Figure 2.2: When light reaches an object with different refractive indices, reflection, absorption and transmission will take place.

2.2 Optical Waveguide

By using an optical waveguide, it is understood to be the guidance of an electromagnetic wave in a suitable structure. The light undergoes a prescribed direction so that it can be guided to reach desired optoelectronic achievements. There are different types of optical waveguides. Here, the planar waveguide and the striped waveguide are concisely presented.

2.2.1 Planar Waveguide

The guidance of an electromagnetic wave in a planar thin-film is based on the principle of total reflection. A planar waveguide consists of optical layers with different refractive indices. The layer, which guides the light, has a refractive index of n_f. It is surrounded by a bottom layer (which is normally a substrate) with a refractive index of n_s and an upper layer (in the simplest case air) with a refractive index of n_c. If $n_s = n_c$, the waveguide is symmetrical, otherwise, it is an asymmetrical planar waveguide.

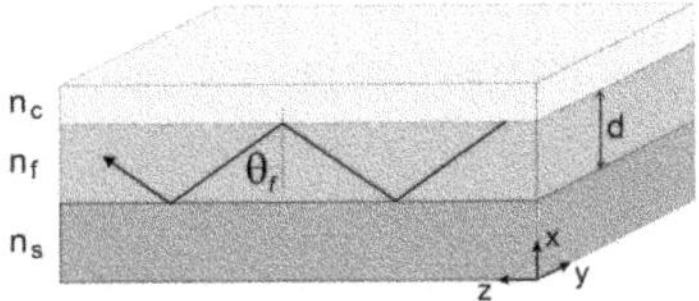

Figure 2.3: Structure of a planar waveguide. Reprinted from [39].

Figure 2.3 shows an example of a stack of layers, which is under the idealized assumption of infinite extension in the y- and z-directions. Transmission of an electromagnetic wave in the guiding layer would be possible if the refractive indices of the layers satisfy the inequality of $n_c \leq n_s \leq n_f$. Light performs a zigzag movement in the z-direction at an angle between the boundary layers. For the critical angle of total reflection in the guiding layer, $\theta_f > \theta_s, \theta_c$, in which θ_s and θ_c are in the relationship with the refractive index of the respective material like the equations below:

$$sin\theta_c = \frac{n_f}{n_c} \ and \ sin\theta_s = \frac{n_f}{n_s}. \tag{2.4}$$

The propagated waves can be distinguished into transverse electrical (TE) and transverse magnetic (TM). In TE waves, the electric field vector is perpendicular to the plane of incidence (x-z direction), whereas the magnetic field vector of TM waves is perpendicular to the direction of propagation. In addition to the condition of total reflection, every guided wave must "fit into" the guiding layer. In the x-direction, the wave components must interfere perpendicularly to the interfaces, to interfere with the standing waves. For this, the reflection must be in phase with x-components of the electromagnetic waves to provoke constructive interference. When these conditions are fulfilled, it is called a glancing angle, which ensures the propagation of discrete waves. This is defined as mode. A wave, which propagates in the z-direction, is described by the phase constant of the propagation coefficient β:

$$\beta = k n_f sin\theta \tag{2.5}$$

whereas k is the wave vector, which can be determined according to:

$$k = \frac{2\pi}{\lambda} = \frac{\omega}{c} \tag{2.6}$$

whereas ω is the angular frequency. For those under the angle of total reflection, the incident waves together with the reflected waves will have an effective refractive index n_{eff} as follow:

$$n_{eff} = n_f sin\theta = \frac{\beta}{k}. \tag{2.7}$$

The effective refractive index is a mode-specific variable, which is determined by all refractive indices. Thus, it depends on all the layers involved in the thin-film. A different effective refractive index applies to each propagation of mode. The relation $n_c \leq n_s \leq n_{eff} \leq n_f$ is generally valid. The following applies to the condition of the in-phase superposition:

$$k d n_f cos\theta_m - \Phi_c - \Phi_s = m\pi \tag{2.8}$$

whereas d is the thickness of the layer, m is the order of the mode and θ_m is the associated propagation angle, Φ_c and Φ_s are the phase shift at the respective interfaces according to the Goos-Hänchen effect [40, 41]. In contrast to the observation of electromagnetic waves as optical rays, the standing waves are not completely reflected at the interfaces. Part of the field intensity penetrates the adjacent layer and drops exponentially. This field portion is named as evanescent field [42]. The angle of the phase difference at the interfaces is calculated differently for TE and TM modes:

$$tan\Phi_{c,s}(TE) = \frac{\sqrt{n_f^2 sin^2\theta - n_{c,s}^2}}{n_f cos\theta} \tag{2.9}$$

$$tan\Phi_{c,s}(TE) = \frac{n_f^2}{n_{c,s}^2} \frac{\sqrt{n_f^2 sin^2\theta - n_{c,s}^2}}{n_f cos\theta} \tag{2.10}$$

d from Equation 2.8 plays an important factor in the existence of the guided light waves in the thin-film. For a certain layer thickness, only a limited number of modes can propagate. The first mode is also called the fundamental mode, which requires a certain minimum thickness ($d_{cut-off}$). In general, the minimum thickness (d_m) for the possible order of TE modes is calculated by using [43]:

$$d_m = \frac{\lambda}{2\pi}(n_f^2 - n_s^2)^{-\frac{1}{2}}(arctan\sqrt{\frac{n_s^2 - n_c^2}{n_f^2 - n_s^2}} + m\pi) \tag{2.11}$$

In Figure 2.4, the minimum thickness of the guiding layer for the fundamental mode TE_0 and the first subsequent TE_1 mode is shown, which is calculated according to Equation 2.11. It is an asymmetrical waveguide, as the refractive indices of the bottom layer (n_s of silicon dioxide = 1.46) and the upper layer (n_c of air = 1) are not the same. It is calculated for the spectral range from 550 nm to 650 nm based on the needs in this research. When the only fundamental mode is capable of propagation (single-mode), the layer thickness must be at least d_0. If the layer thickness exceeds the value d_1, TE_1 mode will be able to propagate [43].

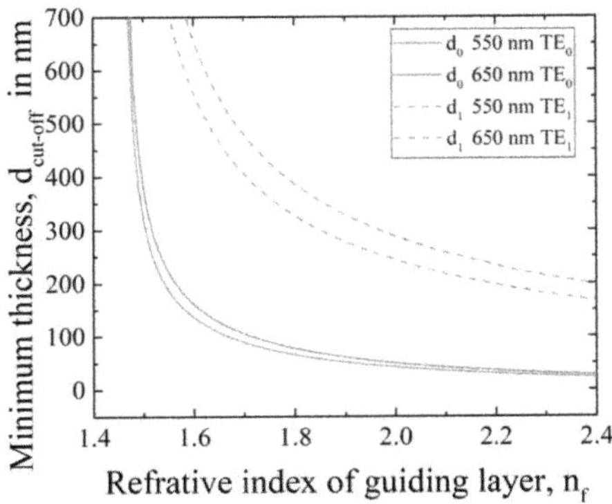

Figure 2.4: Minimum layer thickness $d_{cut-off}$ for TE_0 and TE_1 modes based on Equation 2.11.

In order to determine the field distribution of E(x) and H(x), the Helmholtz equations need to be solved. A simulation program based on Mathematica® was applied for the calculations in this work, which was developed by Rabe et. al. in his dissertation [44]. The fill factor (Γ_j), in which the wave intensity in a layer was distributed, can be calculated by using:

$$\Gamma_j = \frac{\int_{x_{j-1}}^{x_j} |E^2|\,dx}{\int_{-\infty}^{\infty} |E^2|\,dx}. \tag{2.12}$$

2.2.2 Stripe Waveguide

Stripe waveguide is different from the planar waveguide, as it has an additional lateral waveguide, which can be produced by having a rectangular structure in the yz-plane. The guiding waveguide is generally named core, while the surrounding material is called cladding. This structure has an impact on the effective refractive index and the propagation of modes in the waveguide, eventually.

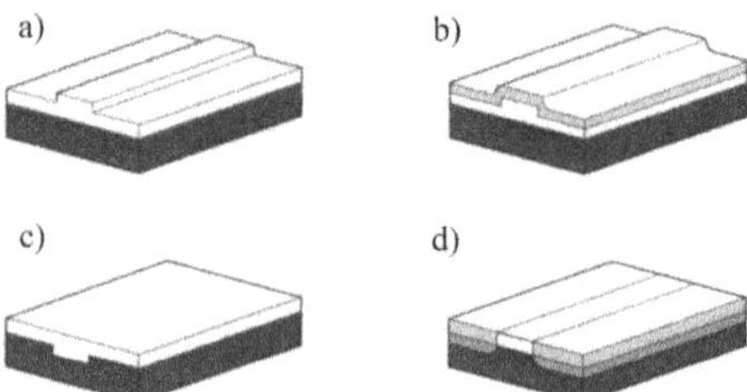

Figure 2.5: Different forms of stripe waveguide are illustrated: (a) ridge, (b) inverted ridge, (c) buried and (d) diffused. Reprinted from [45].

In Figure 2.5, there are some options for realising a stripe waveguide. The ridge waveguide can be produced when the thickness of the guiding material in the middle area of the stripe waveguide is increased (see Figure 2.5(a)). As in Figure 2.5 (b), when the ridge waveguide corresponds in the direction of the substrate or cladding, an inverted ridge waveguide is presented. If the stripe waveguide is sheathed from above as Figure 2.5 (c) and (d), it is referred to as a buried or diffused waveguide. In such a way, single-mode operations can take place with a relatively larger height and width. An advantage can be found in a diffusion waveguide. As the field distribution in the core can be cylinder-symmetrical, a high coupling efficiency is possible in single-mode operations with suitable adaptation. [45, 46].

2.3 Organic Semiconductor

Organic semiconductors can be applied not only for light guiding but also for light generation. Generally, they consist mainly of carbon atoms and hydrogen atoms. Besides, they are also materials with π-bonded molecules, which function as building blocks. The material can be categorised into two groups according to their molecular weight. One of the groups is considered as small molecules when the molecular weight is less than $1000\,\mathrm{g\,mol^{-1}}$. Meanwhile, polymers are with a molecular weight greater than $1000\,\mathrm{g\,mol^{-1}}$. Organic semiconductors are very useful. When small molecules are used, charge transport can be facilely controlled by modifying various molecular parameters, while polymers can be processed to make thin films with large areas conveniently [47].

The essential component of organic semiconductors is the carbon atom. A carbon atom has a basic state of an electron configuration of $1s^2 2s^2 2p^2$. There are two electrons in $1s^2$ state, which are closer to the core. Compared to the remaining four so-called valence electrons in $2s^2$ and $2p^2$, the two electrons in $1s^2$ state are more strongly bounded. When a $2s$ electron is excited with enough energy, it can be raised to the higher energy state $2p$. Benzene is a typical instance of a small organic molecule with the structure described above. In a benzene molecule, there are in total only six carbon atoms (see Figure 2.6).

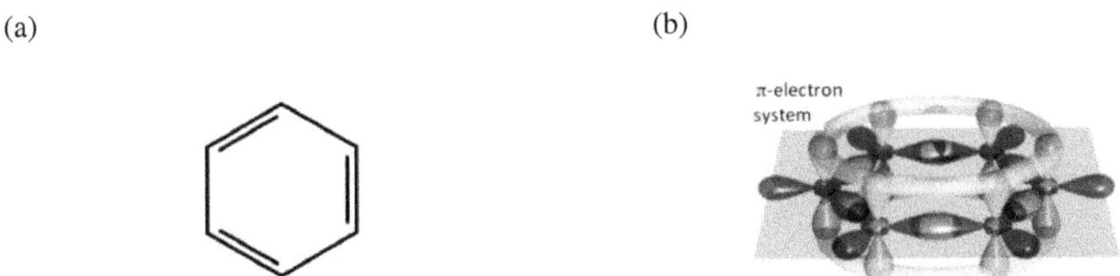

Figure 2.6: (a) The chemical structure and (b) the orbital structure of the small molecule, benzene. Represented from [48]

Due to hybridisation, σ bonds can be formed with the electrons in sp^2 state. Meanwhile, the remaining electrons in p_z orbital form π bonds with the neighbouring p_z orbital. In this way, π-electrons are actually delocalised over the whole molecule. By using the linear combination of atomic orbitals (LCAO), the molecular orbitals of organic molecules can actually be predicted. However, it is not the only possibility. Based on the Aufbau principle, linear combinations of N linearly independent atomic orbitals provide N linearly molecular orbitals. In the example of benzene, six $2p_z$ carbon orbitals induce six molecular orbitals as seen in Figure 2.7. As a result of the distribution of the electron in the energy levels, bonding molecular orbitals (π orbitals) and antibonding molecular orbitals (π* orbitals) exist. There is the highest occupied molecular orbital (HOMO) in the bonding molecular orbitals and the lowest unoccupied molecular orbital (LUMO) in the antibonding molecular orbitals [49].

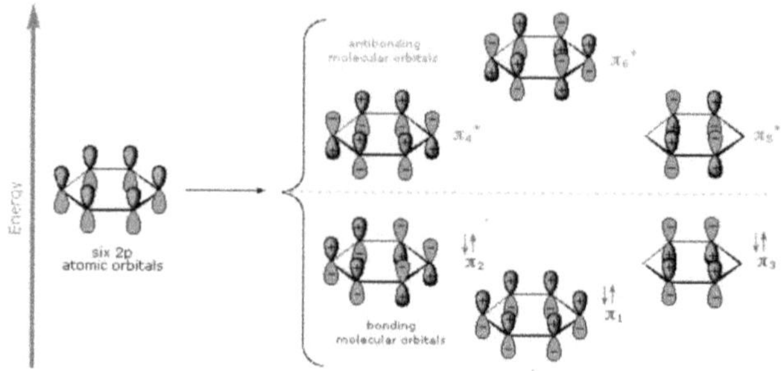

Figure 2.7: Due to the electron distribution, bonding and antibonding molecular orbitals are formed, which lead to HOMO and LUMO levels. Reprinted from [50].

The conjugation of bonds generates electron delocalisation. This contributes to the semiconducting properties of organic materials. The difference in the energy levels between HOMO and LUMO is the so-called energy gap of organic semiconductors, which is one of the important properties of semiconductors:

$$E_{\text{gap}} = E_{\text{LUMO}} - E_{\text{HOMO}} \tag{2.13}$$

The energy gap of organic semiconductors has typical values in the range from 1.5 eV to 3.5 eV [49]. It depends strongly on the semiconductor properties such as optical absorption and emission process of the molecule. Meanwhile, the energy gap can be tuned by composing the molecule in the desired way because the magnitude of the energy gap is related to the specific molecular structure [51]. This provides a vast variety to obtain different organic compounds. Thus, designing of organic materials becomes very important, so that the specific absorption and emission properties can be realised.

The application of organic semiconductors as light sources relies on the energy gap between the excited states and the ground state. Whereas, this can correspond to the range of visible light. The first step to have light emission is the excitation of the electrons of the molecules. There are two of the most common methods of excitations, which are electrical excitation (electroluminescence) and optical excitation (photoluminescence). The light forming process depends mainly on hole-electron pair, which is also called excitons constructed by Coulomb interaction [52].

In this work, to excite the samples, photoluminescence takes place to derive light from the organic materials. The molecules are optically excited. Whereas, the excited states of the molecules can be generally classified into two groups: singlet states with a total spin of 0 and triplet states with a total spin of 1. When there is radiative relaxation from the singlet states to the ground states, it is named fluorescence, while the mechanism from triplet states to ground states, it is called phosphorescence. This can be elucidated by using the Jablonski diagram (see Figure 2.8). Intersystem crossing can take place between the singlet states and the triplet states when the energy gap between them is small enough. In this process, excitons can transit from the triplet states to the singlet states and from the singlet states radiatively relax to ground states, eventually. The whole mechanism is named thermally activated delayed fluorescence (TADF) [53].

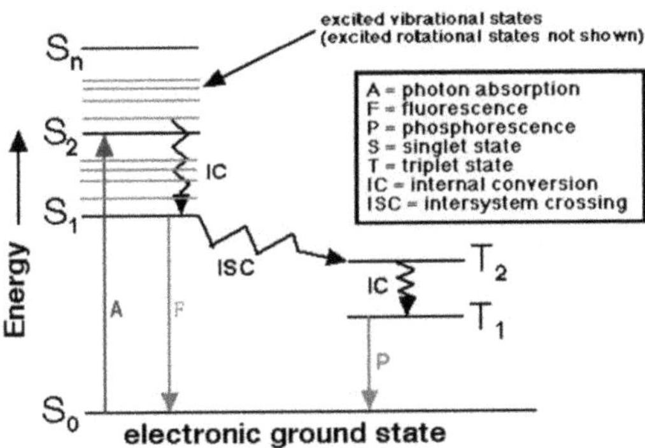

Figure 2.8: Through Jablonski diagram, luminescence is described. Reprinted from [54].

2.3.1 Photoluminescence Quantum Yield (PLQY) and Fluorescence Lifetime

The PLQY of organic materials is probably one of the most important characteristics of a fluorophore. It is defined as the number of emitted photons relative to the number of absorbed photons. When the substances have a higher PLQY, they can display brighter emissions. For the determination of PLQY, the emissive rate of the fluorophore (Γ_f) and its rate of non-radiative decay to the ground state (k_{nr}) are taken into consideration. Both of these constants depopulate the excited state. These can be explained by a simplified Jablonski diagram (s. Figure 2.9).

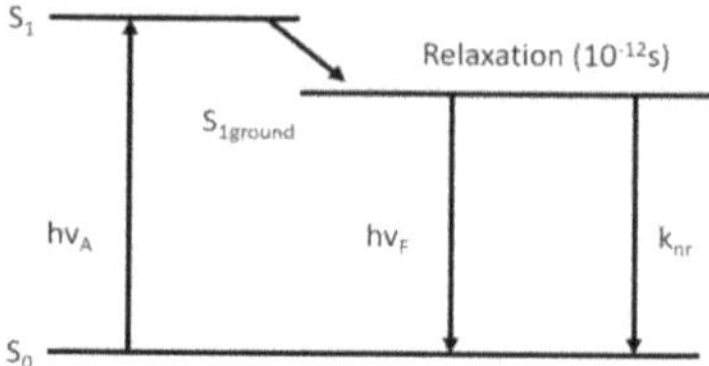

Figure 2.9: A simplified Jablonski diagram is used for the description of quantum yield. Reprinted from [55].

The PLQY can also be illustrated by using the equation below:

$$PLQY = \frac{\Gamma_f}{\Gamma_f + k_{nr}} \tag{2.14}$$

If the non-radiative decay is much smaller than the rate of radiative decay ($k_{nr} < \Gamma_f$), the PLQY can be close to unity. However, it is always not the case. The PLQY is less than unity due to the Stokes losses. Two similar molecules can have different emission spectra and quantum yields due to the differences in non-radiative decay rates, for example: eosin and erythrosin B. Meanwhile, heavy atoms can induce a lower quantum yield.

Other than the PLQY, the fluorescence lifetime of a fluorophore is another factor that cannot be neglected. It is determined by the average duration of the excited molecule in the excited state before returning to the ground state. This is because fluorescence emission is a random process, while only a few molecules emit their photons at a precise period. When a single exponential decay takes place, 63% of the molecules have decayed before the fluorescence life time (τ_{fl}). When the time is > τ_{fl}, the rest 37% decay. According to the illustration in Figure 2.9, the τ_{fl} can be described through the following equation:

$$\tau_{fl} = \frac{1}{\Gamma_f + k_{nr}} \tag{2.15}$$

In the absence of non-radiative mechanisms, the lifetime of the fluorophore is also named the intrinsic or natural lifetime, while the equation is shown as:

$$\tau_{fl} = \frac{1}{\Gamma_f} \tag{2.16}$$

2.3.2 Quenching

The intensity of fluorescence can be minimised by different sorts of processes. This kind of mechanism is named quenching. There is a variety of reasons of quenching. The excited fluorophore can be deactivated when it is in contact with some other molecule in the solution, which is defined as the quencher. In this case, the fluorophore needs to undergo a diffusive encounter with the quencher, in order to return to the ground state, while the molecules are not chemically altered in the process. The decrease in intensity can be illustrated by the Stern-Volmer equation:

$$\frac{F_0}{F} = 1 + K[Q] = 1 + k_q \tau_0 [Q] \tag{2.17}$$

whereas, F is the fluorescence, K is the Stern-Volmer quenching constant, k_q is the bimolecular quenching constant, τ_0 denotes the unquenched lifetime and $[Q]$ denotes the concentration of quencher. The sensitivity of the fluorophore to a quencher is demonstrated by the Stern-Volmer quenching constant. When a fluorophore is buried in a macromolecule surrounding, the value of K is low because it is usually inaccessible to water-soluble quenchers. On the other hand, K becomes larger, when the fluorophore is free in solution. The process of quenching is varied with the fluorophore-quencher pair. For instance, quenching caused by halogen or heavy atoms takes place as a result of spin-orbit coupling and intersystem crossing to the triplet state.

Besides, fluorophores can become non-fluorescent complexes with quenchers under static quenching. It occurs in the ground state and does not depend on diffusion or molecular collisions. Moreover, quenching can also be provoked by numerous trivialities, which could be non-molecular mechanisms, such as attenuation of the incident light by the fluorophore itself or other absorbing species.

3 Organic Laser

In this work, the feasibility of dye-doped PMMA as an organic dye laser was tested. Therefore, it is essential to understand the basic knowledge of organic lasers. In this chapter, the working principle and the device structure of a distributed feedback (DFB) laser are briefly introduced. In addition, a general literature overview is presented, so that the state-of-the-art of thin-film lasing can be understood. Indicated references are provided if there is interest for detailed information.

3.1 Working Principle

The fluorescence mechanisms described in the previous section are named spontaneous emission. On top of absorption and spontaneous emission, stimulated emission is another kind of interaction between light and matter, which can be triggered by an electromagnetic wave. The resulting photons consist of the same properties in phase and frequency. In addition to this, the propagation of light takes place in the same direction. Whereas, stimulated emission is the elemental basis for realising a light amplification by stimulated emission of radiation (laser). Figure 3.1 shows the principal structure of a laser. Three components are needed: excitation source, resonator, and active medium. Fundamentally, a suitable active medium is required, in which the emission is reflected into the active medium for light amplification using an optical structure called a resonator. The two mirrors shown in Figure 3.1 function as resonators to amplify the light, while one of the mirrors does not possess 100 % reflectivity, so that part of the laser light can be extracted.

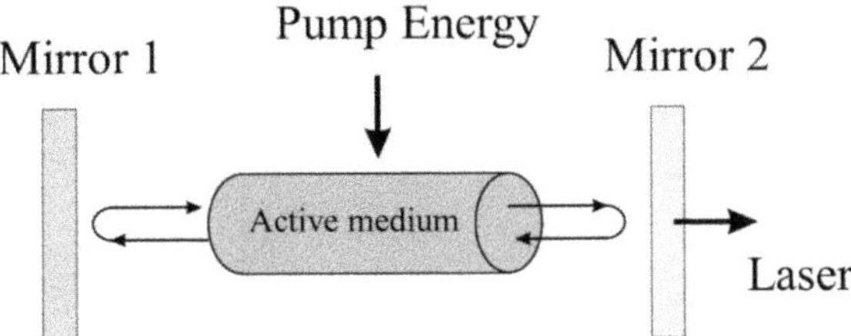

Figure 3.1: Basic structure of a laser. Reprinted from [58].

To achieve light amplification, a mechanism is needed, which is defined as population inversion, where more excited electrons are in higher excited states than the ground state. Therefore, a 2-level system is not suitable to generate a laser because it is always in a condition of thermal equilibrium. Laser generation may be attainable by using a 3-level or a 4-level system. The physical processes like excitation, decay, and laser transition of these systems are illustrated in Figure 3.2 [60].

In a 3-level system, certain numbers of electrons will be excited to level 3 upon energy pumping. To have laser operation, these excited electrons must decay to level 2 quickly. The growing rate of the excited electrons at level 2 must be higher than the rate of spontaneous decay of the pumped level 3 back to the ground state (level 1). In this case, population inversion can be achieved as there are more excited electrons in level 2 than in level 1 ($N_2>N_1$). Subsequently, laser transition occurs when the excited electrons "relax" from level 2 to level 1. To achieve population inversion, at least half of the population of the electrons of the ground states must be excited, which results in an inefficiency of a 3-level laser [60].

A 4-level-laser is one of the typical patterns of energy level for lasers. Just as the name suggests, there are four energy states involved in the laser process. Organic laser is considered as quasi-four-level laser, while the individual transitions include singlet states. The mechanism is described in Figure 3.2 (b). At first, by absorption upon certain energy, there is a transition from the ground states (level 4) to a higher vibration level of the first singlet state, S1 (level 3). Then, the excited electrons relax from there without radiation in the lowest vibration level of S1 (level 2). The laser transition takes place from this level in the form of stimulated emission into a vibration level of the ground state (level 1). The final transition from level 1 to level 4 again takes place without radiation by relaxation, in which the relaxation process is very fast (in the order of magnitude of 10^{-13} s [59].) It is even faster than the decay time of the laser transition. In such a way, it is possible to achieve a population inversion $N_2 > N_1$. A 4-level system provides the benefit of low threshold lasing because no minimum pump power is needed for population inversion since the lower level of the laser transition is above the ground state [60, 61].

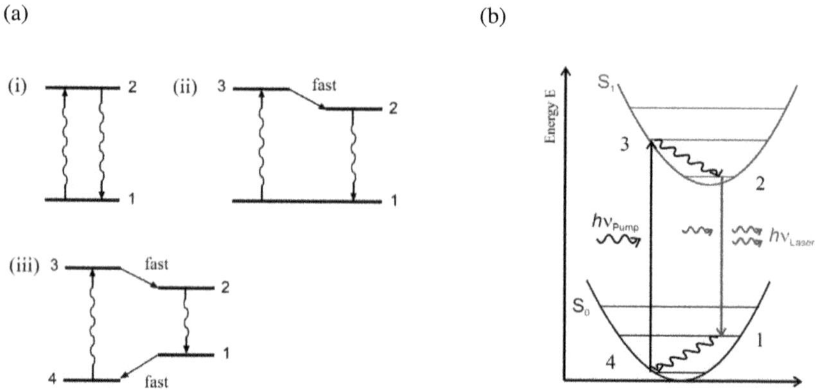

Figure 3.2: (a) The schemes of a 2-level system (i), a 3-level system (ii) and a 4 level system (iii) are illustrated. (b) Energy levels with singlet states in a 4-level system are presented. Represented from [60, 61]

However, an influential aspect that should not be neglected is the presence of triplet states (see Figure 3.3). Intersystem crossing (ISC) may take place, where the molecules in the first excited state S_1 may end up in the T_1 triplet states. Due to the low non-radiative decay (k_T) rate from T_1 to S_0, it may start to accumulate until T_1 goes to higher states like T_2 (triplet absorption). Triplet absorption could be very broad, which could overlap the singlet fluorescence emission band. On top of this, stimulated emission could also be hindered through a non-radiative energy transfer between a molecule in S_1 and another in T_1 (Singlet-Triplet Annihilation). Therefore, OSSLs can only be excited or radiated by short pulses. On the other hand, liquid dye lasers can operate in continuous wave mode. This is because it provides a continuous circulation of the fluid, which allows permanent refreshing of the medium. Efforts have been made to electrically pump an organic laser, but it is not successful due to the following reasons: Firstly, the application of electrodes for electrical injection is troublesome. The guide mode may slip outside the high-index active part. Eventually, it may overlap with the absorbing metallic electrodes. As a consequence, the threshold is enormously raised [62]. Secondly, the problematic source with electrical pumping is the existence of charge carriers (polarons). They have a wide absorption band, which overlaps the emission band and is able to absorb laser photons and quench singlet excitons, eventually [63]. Thirdly, which is also the biggest obstacle, there is the presence of triplet excitons. Under electrical excitation, triplets are more abundant than singlet excitons, which correspond to a probability ratio of 3:1. As mentioned before, triplets will cause triplet absorption and singlet-triplet annihilation, which will absorb laser photons [64].

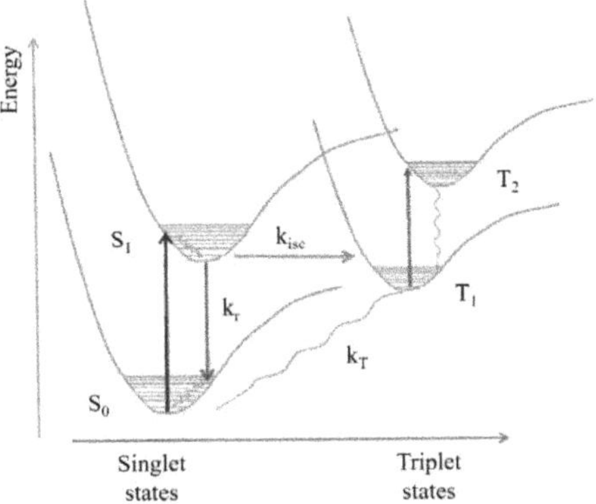

Figure 3.3: Energy level diagram for organic dyes which includes singlet states and triplet states is displayed. Reprinted from [64].

Since there are so many problems with the application of electrical excitation, optically pumping is commonly used for the realisation of organic lasers. The organic material can be excited by lasers, LED, and so on. Therefore, it is important to know the absorption spectrum of the organic material, in order to find out whether the sample can be excited by a certain wavelength or which wavelength is suitable for the optical excitation. By using the absorption spectrum, absorbance and transmittance can be calculated using the Beer-Lambert law, proposed by Pierre Bouguer [65], The relationship of the incident light (I_{inc}) on a film thickness (t) can be described by the following equation:

$$I_{trans} = I_{inc} exp(-\alpha_{abs} t) \tag{3.1}$$

whereas I_{trans} is the transmitted light and α_{abs} is the absorption coefficient. The absorption coefficient, α_{abs} is a scale representing the strength of absorption of the pumped wavelength. Based on Equation 3.1, it can be reformed as:

$$\alpha_{abs} = 2.303(A/t) \tag{3.2}$$

whereas A is the absorbance in the unit of optical density.

Deviation from Beer-Lambert law can arise from instrumental and intrinsic causes. In certain samples, there are macromolecules or large aggregates which induce light scattering. The absorbance caused by scattering is proportional to $1/\lambda^4$ based on Rayleigh scattering, which can probably be recognised as a background absorption that increases rapidly with decreasing wavelength. Besides, there is another deviation: if the samples are absorbing species that present fluorescence simultaneously, the emitted light should not be detected by the detector. This problem can be minimised by setting the detector distant from the samples so that the fluorescent light of the samples is hardly collected. Moreover, another deviation can occur, when the concentration of the samples is too high, or the absorbing species are only partially soluble in the samples. Aggregates will develop when the concentration increases. The absorption spectra of the aggregates and the monomer of the samples may not be the same. For example, the dye bromophenol appears to be blue but it turns red when the concentration rises. Furthermore, instrumental artifacts can provoke absorbance that is nonlinear with concentration, which generally happens when the concentration is too high [55]. Therefore, one needs to take considerable precautions to reduce the unnecessary deviations.

The process of stimulated emission is not only due to the presence of excited states but also to the photon density in the material. It can be illustrated by optical amplification, yet be counteracted by absorption.

$$I(\lambda) = I_0(\lambda)exp[(g(\lambda) - \alpha(\lambda))L] \tag{3.3}$$

whereas I_0 denotes the incident intensity and L indicates the length of the path covered in the medium. The optical gain (g) is determined by the product of the cross section of the stimulated emission (σ_{SE}) and the density of excited states $N_{(2)}$. Generally, the spectrum of optical gain is very similar to the photoluminescence spectrum. Therefore, it can be linked via σ_{SE} by using the equation below:

$$g(\lambda) = \sigma_{\mathrm{SE}} N_{(2)} \tag{3.4}$$

$$\sigma_{\mathrm{SE}} = \frac{\lambda^4 \tilde{I}_{\mathrm{D}}(\tilde{\lambda})}{8\pi n^2 c \tau_{fl}} \tag{3.5}$$

whereas $\tilde{I}_{\mathrm{D}}(\tilde{\lambda})$ is the standardised emission spectrum, n is the refractive index, c is the speed of light in a vacuum, τ_{fl} is the fluorescence lifetime.

Meanwhile, the Beer-Lambert law together with the pump energy can used to compute $N_{(2)}$ based on the equation below:

$$N_{(2)} = \frac{(\frac{E_{in}}{\lambda_{exc}})PLQYexp(-\alpha_{abs}t)}{V_{sam}} \tag{3.6}$$

whereas E_{in} is the excitation energy in the unit of joule, λ_{exc} is the excitation wavelength in the unit of joule, t is the thickness of the sample and V_{sam} is the volume of the sample.

The stimulated emission cross section is an important factor, which has an influence on the lasing properties. For example, a figures-of-merit (FoM) is proposed by Miniscalco et al. based on Nd^{3+}-doped fiber amplifier. [66].

$$FoM = \frac{1}{L_{th}} = \frac{4.34}{hv_pA_{eff}}[\sigma_{\mathrm{SE}}(v_s) - \sigma_{\mathrm{esa}}(v_s)]\tau_{fl} \tag{3.7}$$

whereas L_{th} is the lasing threshold, v_p is the photon frequency of pump wavelength, A_{eff} is the overlap integral between the signal mode and the uniformly doped core, v_s is the photon frequency of signal wavelength and σ_{esa} is the cross section of strong excited state absorption transition. Based on the theory, this FoM is a reciprocal representative of the lasing threshold. Therefore, Equation 3.7 shows that a higher stimulated emission cross section induces a lower lasing threshold.

3.1.1 Amplified Spontaneous Emission (ASE)

When the stimulated emission predominates the spontaneous emission, the mechanism is called amplified stimulated emission (ASE). Even though the origin of ASE is the statistically undirected processes of the spontaneous emission, which induces the directionality of the individual stimulated emission process. ASE has a length-dependent change in the spectral composition and is angle-dependant (see Figure 3.4).

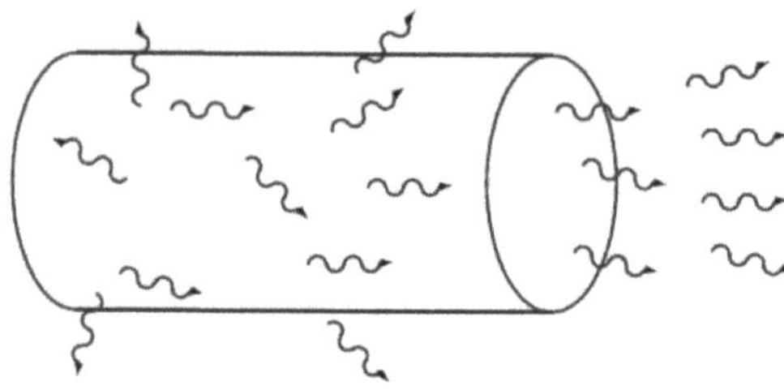

Figure 3.4: The directional characteristics of amplified spontaneous emission. Reprinted from [67].

The macroscopic gain of the radiation intensity is directionally the strongest, in which the distance covered by the excited light is the longest. The factor of the length-dependence in the intensity is shown through the equation below [68, 69]:

$$I(L,\lambda) = \frac{I_{sp}(\lambda)}{g(\lambda)}(exp(g(\lambda)L) - 1) \tag{3.8}$$

whereas $I_{sp}(\lambda)$ denotes the intensity of the spontaneous emission, $g(\lambda)$ is the optical gain and L is the excited length. The intensity increases exponentially with a positive optical gain. This behaviour is used to determine the spectral optical gain by using the varied stripe length method (VSL). The optical gain adjusts itself according to the following equation:

$$g(\lambda) = \Gamma g(\lambda) - \alpha_W(\lambda) \tag{3.9}$$

whereas Γ is the fill factor and $\alpha_W(\lambda)$ is the optical loss. The material gain is determined by weighting the fill factor because the optical wave is amplified only in the active area. The fill factor is calculated as the part of the total intensity that runs in the active zone [70] based on Equation 2.12.

In addition to the changes of the directional characteristic, the spectrum of ASE is also different. The spectral components are stimulated more frequently in the region of gain maximum (g_{max}). The emission process is disproportionately strengthened, which causes a constriction of the emission spectrum at the frequency of g_{max}. The relationship is expressed quantitatively as follows:

$$\frac{I_v}{I_{max}} = \frac{exp(g_v L) - 1}{exp(g_{max} L) - 1} \tag{3.10}$$

whereas I_{max} is the intensity maximum of ASE at the frequency v and g_v is the gain coefficient of the material at the frequency v.

Figure 3.5 shows a spectrum of Gaussian emission for comparison. There are two spectra of normalised ASE intensity with two different values of gain $G = gL$. It can be clearly recognised that there is a narrowing of the spectrum with increasing gain. For calculation, the gain spectrum is assumed to be a Gaussian distribution. The relationship of the FWHM of ASE spectrum (Δv_{ASE}) to the FWHM of the spontaneous emission (Δv_0) is shown through the equation below:

$$\frac{\Delta v_{ASE}}{\Delta v_0} = \left(\frac{exp(gL) - 1}{gLexp(gL)} \right)^{\frac{1}{2}} \tag{3.11}$$

The relation in Equation 3.11 is illustrated in Figure 3.5(b). In general, it shows the process from stimulated recombination process to a narrowing of the spectrum. This mechanism happens only when $g \cdot L > 1$. This value is often referred to as the threshold value for ASE. However, the ASE threshold cannot be obviously identified. For example, the transition from spontaneous emission to ASE in the curve in Figure 3.5(b) is weakly pronounced. Furthermore, ASE emission is not necessarily coherent due to its origin of spontaneously emitted photons. Nevertheless, based on the spectral analogies, the ASE process is often referred to as mirrorless lasing, in which there is no feedback of light due to the lack of a resonator [71–73]. Fundamentally, ASE is considered as a piece of essential evidence, if the material can practice lasing activity.

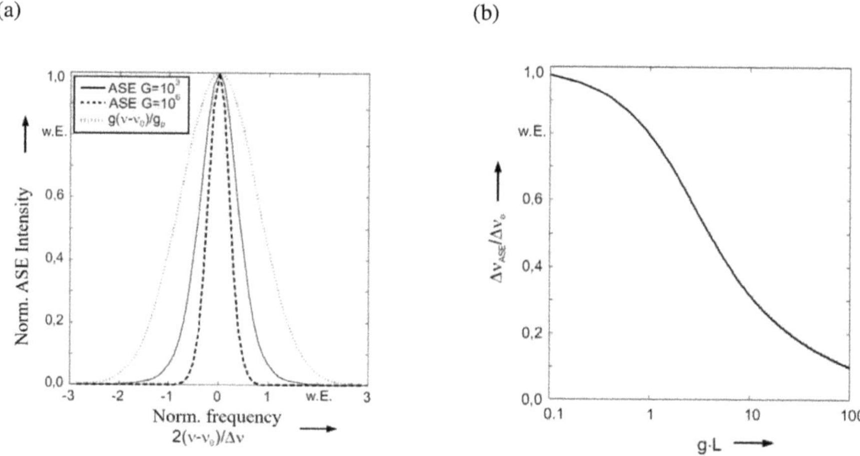

Figure 3.5: (a) Normalised spectrum of ASE at two different excitation densities compared to the spectrum of Gaussian Emission. (b) The development of Full-Width-Half-Maximum as a function of the product of g and L. Reprinted from [74, 75]

3.1.2 Resonator

In addition to the active material, resonator is needed to produce laser light. The simplest construction of a laser is shown in Figure 3.1, in which two parallel mirrors are used for the optical feedback. This kind of structure is called a Fabry-Perot resonator. Standing waves are formed based on the following equation:

$$L_r = m\frac{\lambda}{2n} \tag{3.12}$$

whereas L_r is the length of the resonator, i.e. the distance between the mirror surfaces, n denotes the refractive index and m represents the mode order [76]. The modes running in optical resonators are defined as resonator modes. If the L_r becomes larger, more modes are capable of propagation.

A large optical gain can be achieved by strong optical feedback if the mirrors have good reflectivity. The reflectivity of the resonators is an important factor of laser quality [77, 78]. Depending on the application or the types of the laser, the reflectivity can also be induced by the reflection at the interface between two materials with different refractive indices. For instance, a Fabry-Perot laser could be realised through the difference of refractive indices between the organic materials ($1.6 < n_{organic} < 2.0$) and the adjacent air ($n_{air} = 1$) [79]. One of the two mirrors used must have a slightly transmitting effect for the photons so that the laser light can be out-coupled. Before a laser can be generated, a large number of photons has to be achieved through stimulated emission. As a result, a laser functions upon a certain working point, which is defined as the lasing threshold. This describes the state, in which the gain through pumping processes just compensates for the losses due to the absorption in the laser-active medium.

Due to the simple processability, organic materials are opened up to different varieties of other possible resonator structures, other than Fabry-Perot resonators. In Figure 3.6 (a), the implementation of planar microresonators is shown, which was first published by Tessler et al. in 1996 [80]. In this approach, a sandwich construction of one thin organic layer between two different mirrors is proposed. A highly reflective dielectric mirror is placed on the bottom, whereas the laser light can pass through a thin metallisation on the other side of the structure. One clear disadvantage of the application of microresonators is the small distance between the resonators, which associates a small optical gain. Another two interesting resonator geometries are illustrated in Figure 3.6 (b) and (c), which include the employment of micro-drops [81, 82] and micro-rings [83, 84]. Under these two constructions, the electromagnetic wave stays in the active material for a long time, which can experience a high level of reinforcement. Moreover, low energy is required to reach the lasing threshold. The resonator shown in Figure 3.6 (d) plays an important role in this research approach. In this structure, an optical grating is used. A thin-film laser with optical grating directly on the gain medium is based on the principle of a distributed feedback of the light. Thus, the laser is called a distributed feedback (DFB) laser, which will be further discussed in the following chapter.

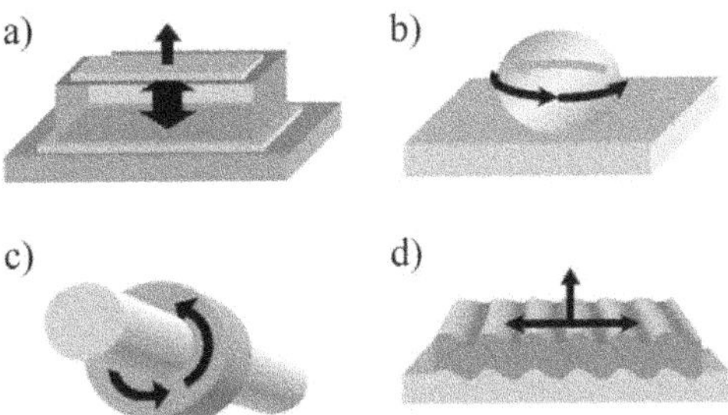

Figure 3.6: Different geometries of optical cavity for organic lasers: (a) planar microresonator, (b) micro-drops, (c) micro-ring and (d) periodic structure. Reprinted from [85].

3.2 Distributed Feedback (DFB) Laser

Single-mode operations in a Fabry-Perot laser are very challenging because there are many resonating longitudinal modes that undergo the same diffraction losses or the same gain in the material. The distance between the resonators is often too large compared to the intended wavelength for oscillation. This induces a small mode spacing, which causes the oscillation of several frequencies at the same time. This phenomenon is also known as the frequency comb [86]. To establish a single-mode operation, the resonator mirrors can be replaced by a spatially periodic grating structure. Such a structure leads to a periodic modulation of the effective refractive index. If the grating structure is passively connected to both sides of a gain medium of a laser, the laser is called Distributed Bragg Reflector (DBR) Laser. On the other hand, it is defined as a Distributed Feedback (DFB) laser if the grating structure is directly built on top of the gain medium of a laser and is used for light amplification.

Some samples prepared in this work are one-dimensional DFB lasers, which operate in the second-order of diffraction. It must satisfy the Bragg condition shown below:

$$\lambda_{Bragg} = \left(\frac{2n_{eff}\Lambda}{m}\right) \tag{3.13}$$

whereas n_{eff} is the effective refractive index, m is the order of diffraction and Λ is the grating period. λ_{Bragg} is the resonant wavelength in the cavity, which is reinforced during its propagation along the active layer, before it is diffracted in the grating in different directions. In a second-order DFB laser ($m = 2$), the light can be coupled out of the film mainly in a direction perpendicular to the active layer by the first-order of diffraction, if it is excited at a certain angle to the normal of the sample surface. According to the coupled wave theory [87], DFB lasers are dominated by index coupling. The emission occurs possibly with a pair wavelength on either edge of a prohibited band and it is centred at λ_{Bragg}. However, it is another case for second-order devices. The peak with a shorter wavelength has a larger threshold due to radiation losses [88]. Correspondingly, the peak with a longer wavelength can be observed, whereby there is phase discontinuity or phase shift found in the devices, which is not necessary [89]. Under correct circumstances, a single-mode emission spectrum is usually found, which consists of only a single peak.

The coupled wave theory starts with one-dimensional Helmholtz-equation:

$$\frac{d^2}{dz^2}E_y(z) + \tilde{k}^2(\omega, z)E_y(z) = 0 \tag{3.14}$$

The complex circular wave number $\tilde{k}$ contains the real part of the refractive index modulation $n(z)$ and the imaginary part of the possible gain or loss modulation g(z). If one restricts the analysis to small variations of the refractive index,

$$\tilde{k}^2(\omega,z) \approx \beta^2 + i\beta g + 4\beta\kappa cos(\frac{2\pi}{\Lambda}z) \tag{3.15}$$

whereas the coefficient of wave propagation $\beta = \omega n_{eff}/c$ and the complex coupling factor $\kappa = \pi\Delta n/\Lambda + i\Delta g/4$. Such an approach represents a linear combination of outgoing and returning waves (in the z-direction), which couples with one another through the periodic structure and energy exchange. The equation seems to be the following:

$$E(z) = E_+ exp(i\beta_b z) + E_- exp(i\beta_b z) \tag{3.16}$$

whereas the phase constant of the grating can be elaborated as $\beta_b = m\pi/\Lambda$. According to [87], terms of second-order are neglected and only the area which is close to the Bragg wavelength λ_b is considered. Thus, the coefficient can be compared:

$$\frac{dE_+}{dz} + \left(\frac{g}{2} - i\delta\right) E_+ = i\kappa E_- \tag{3.17}$$

$$\frac{dE_-}{dz} - \left(\frac{g}{2} - i\delta\right) E_- = -i\kappa E_+ \tag{3.18}$$

whereas $\delta = \beta - \beta_b$ denotes the detuning factor of the wavelength from the Bragg wavelength. The coupling equations describe the wave propagation through two opposite coupled modes. The strength of the interaction is given by the coupling factor. A dispersion relation can be derived from the differential equations above, while the boundary at the end faces of the resonator is taken into account. It defines the wavelength and the gain of the resonator modes. For a grating of first order with length, L and perfectly reflective end surfaces, some analytical approximate solutions to the equations above are given from [87, 90].

In the case of pure refractive index modulation with $\Delta g = 0$ and no natural oscillation, there is a mode spectrum that is symmetrical to the Bragg wavelength. A so-called stop band with the width

$$\Delta\lambda_{stop} \approxeq \frac{\kappa\lambda_b^2}{\pi n_{eff}} \tag{3.19}$$

is created, where no wave propagation is allowed, or strong attenuation occurs. The distance of the modes outside the stop band corresponds to the mode distance in the Fabry-Perot resonator.

$$\Delta\lambda_{DFB} = \frac{\lambda^2}{2n_{eff}L} \tag{3.20}$$

As the gain profile of the active medium is symmetrical to the Bragg wavelength, two modes with a minimum threshold can oscillate at both edges of the stop band. This can be calculated by using the equation below:

$$g_{th} \approxeq 2\frac{\pi^2}{\kappa^2 L^3}. \tag{3.21}$$

The mode distribution as a function of the gain threshold to the wavelength mode is shown in Figure 3.7. No stop band is formed in the case of pure gain modulation because no interference can arise from reflection due to changes in the refractive index. Mode selection is obtained by overlapping longitudinal modes with the reinforcement grating. The mode with the lowest gain threshold value occurs at λ_B. The greater the distance from the λ_B, the greater the increase in the gain threshold [87, 91].

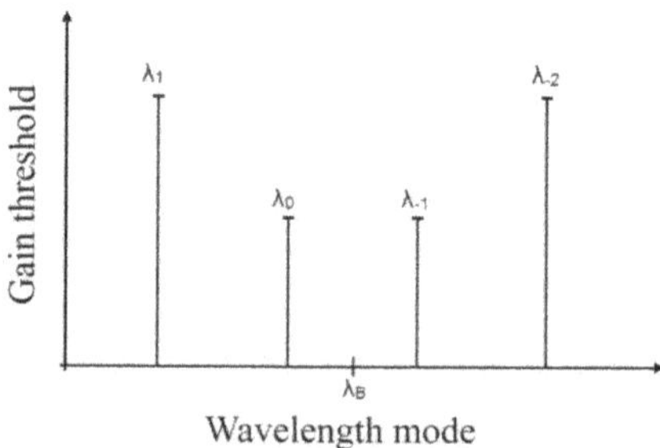

Figure 3.7: Gain threshold of the modes in a DFB laser. Reprinted from [91].

3.3 Literature Overview

There are several methods to achieve a polymer thin-film laser. Here, the state-of-the-art of organic dye-doped polymer thin-film laser is briefly discussed. One of the simplest methods is the direct implementation of thin-films which is supported by the theory of Fabry-Perot [92, 93]. On top of that, a thin-film laser can be achieved by fabrication of thin-film combined with resonator components [72, 94, 95]. By applying a grating structure into the film, it can be a DFB polymer thin-film laser. Generally, there are two methods to produce a DFB polymer thin-film laser. Firstly, a grating structure can directly be written into the polymer/gain medium by using electron-beam lithography or holographic lithography [96–102]. Secondly, the gain material can either be solution-processed (e.g.: spin-coating and dip-coating) or vacuum deposited on the grating structure [14, 15, 29, 103–112]. The lasing wavelength of a DFB thin-film laser can be easily tuned by electrically controlling [96], manipulating of the thickness of the active layer [29, 95], stretching of the active film [30, 100, 113, 114], designing of a wedge-shaped film [102, 106], applying of different grating periods [104, 115] and liquid crystal [116, 117].

In literature, there are huge developments of organic gain media. Moses et. al. demonstrated a very first soluble conduction polymer poly [2-methoxy,5-(2'ethyl-hexyloxy) -*p* -pheynlene- vinylene] (MEH-PPV) at the beginning of the nineties, in which the performance of MEH-PPV in solution as laser dye was comparable to that of Rh6G in solution under identical conditions [118]. Since then, thin-film polymers were actively researched. McGhee et al. managed to spin-caste BuEH-PPV to produce a uniform film in the range from 150 to 300 nm. With mere excitation of film samples, the FWHM of the spectrum was only 8 nm at a pump intensity of $1.5\,\mathrm{kW/cm^2}$ [92]. In addition, Deshpande et al. embedded RhB in polyacrylic acid (PAA) films, in which the lasing efficiency was also comparable to Rh6G in methanol. The spectrum of the film consisted of a peak wavelength of 639.5 nm with a FWHM of 10 nm [93]. This showed a possibility of application as OSSLs. With the implementation of a resonator, the FWHM could be further improved. In the research of Persano et al., poly[(9,9-dioctylfluorenyl-2,7-diyl)-co-(1,4-diphenylene-vinylene- 2-methoxy-5-2-ethylhexyloxy-benzene)] was spin-casted on a bottom mirror and a top distributed Bragg reflector was finally evaporated directly on the polymer film. In such a micro-cavity, the spectrum was narrowed from 20 nm below the threshold to less than 2 nm above the threshold [119]. Examples of external resonator setups were a degenerate two-wave mixing (DTWM) experimental set-up [94] and a hemispherical setup [72]. Voss et al. used a simple prism setup, in which a RhB doped poly-methylmethacrylate (PMMA) was attached to the prism. In this setup, there would be two parts of the excited beam on the doped film, which induced an interference pattern and acted as distributed feedback structure of the laser, eventually. The wavelength could be tuned over a range of 30 nm by rotating the prism. Besides, laser tuning as fine as 1 nm could be achieved in [95] (see Figure 3.8).

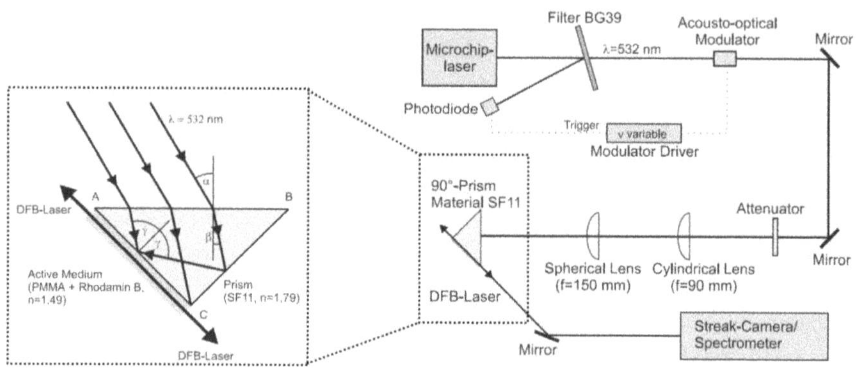

Figure 3.8: The prism setup from [95].

The gain material of DFB laser can be categorised into two groups. First, the polymer itself is a fluorophore. In literature, there were polymers that directly acted as gain materials such as poly[2,5-bis(2',5'-bis(2"-ethylhexyloxy)phenyl)-p-phenylene vinylene] (BBEHP-PPV) [15, 103], poly[phenylene vinylene] derivative [101], poly[(9,9-dioctylfluorenyl-2,7-diyl)-alt-co-(1,4-benzo-2,1',3-thiadiazole)] [120], poly [2-methoxy-5-(2-ethylhexyloxy)-1,4-phenylene vinylene] (MEH-PPV) [96], tris(trifluorene) truxenes (T3) [108], poly[2-butyl-5-(2'-ethyl-hexyl)-1,4-phenylene vinylene] (BuEH-PPV) [92], poly[9,9-dioctylfluorene-co- benzothiadiazole] (F8BT) [100, 109, 113], poly[9,9-dioctyfluorene] (PFO) [110] and poly[phenylene vinylene] [72] and so on. These materials were usually self-invented and cannot easily be found from the market. There is another option, in which dyes were mixed with polymer. The most common used was 4-(Dicyanmethylen)-6-(4-dimethylaminostyryl)-2-methyl- 4H-pyran (DCM2) doped in tris(8-hydroxyquinolinato) aluminium (Alq_3) [104, 105]. This combination was interesting because Alq_3 not only functioned as a matrix to DCM2, but also had the character as a dye which was typically used in OLEDs [121, 122]. Meanwhile, Cehovski et al. was able to integrate the emission of Alq_3:DCM2 into a single-mode polymer waveguide (see Figure **??**).

On top of this, there are also researches that show that dyes can be doped into polymers. The polymers do not exhibit emission and function as a matrix for the dyes. The examples of dyes seen in the literature were: carbon-bridged oligo(p-phenylenevinylene) (COPV) [30], 4-dicyanomethylene-

2-methyl-6-p-dimethyl-aminostyryl-4H-pyran (DCM) [97], Rhodamine 590 dye [107], p-bis(o-methylstyryl)-benzene (Bis-MSB) [112], Pyrromethene 567 [114] perylenediimide (PDI) derivatives [14, 29, 30, 32] and etc. In the meantime, Bragg gratings could be written into polymers like polystyrene (PS) [14, 29–32], poly[vinylalcohol] (PVA) [112, 114], poly[vinyl alcohol] (DCPVA) [97] or even negative photoresist dichromated gelatine (DCG) [30, 32, 97]. Another easier method was spin-coating or dip-coating the gain materials together with the polymer matrix directly on the substrate with Bragg grating [92, 97, 98, 107–110]. Zhai et al. presented another new structure design of DFB thin-film laser [120]. The grating structure was fabricated separately on top of a polymer film, while a layer of gain material was beneath it. Based on his theoretical model, the field distribution of the eigenmode was confined almost thoroughly in the active layer (see Figure 3.9 (b)). However, Figure 3.9 (a) illustrates that, by assembling the gain medium on top of the grating structure, the eigenmodes were distributed partly in the grating structures and partly in the active layer. Therefore, by using the design in Figure 3.9 (b), the active volume could be used more effectively. Furthermore, the pump energy could be more efficiently extracted. Quintana et al. compared the structure of Zhai with another two sample structures [30]. It was successfully proven in the research that the structure of Zhai provided a lower lasing threshold and a better laser slope efficiency. On top of this, the grating depth and the residual layer of the grating structure were manipulated, while the result is shown in Figure 3.10. In [32], when the thickness of the residual layer decreased and the grating depth increased, the peak wavelength shifted slightly to a shorter wavelength. Besides, the relationship between the grating depth and the laser efficiency is drawn in Figure 3.10 (b). When the grating depth increased, the laser efficiency was improved as well.

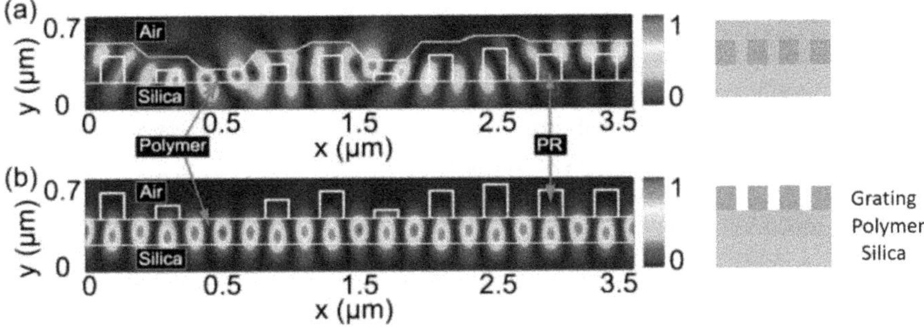

Figure 3.9: Structure comparison by Zhai et al. in [120]. Redrawn from [120].

(a) (b)

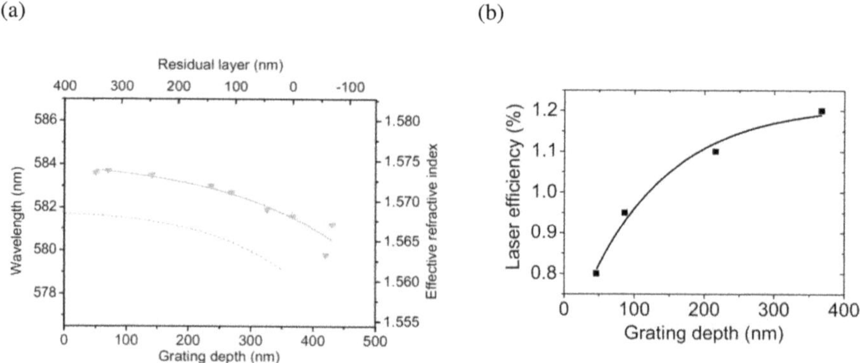

Figure 3.10: (a) Relationship between grating depth, residual layer, wavelength and effective refractive index. (b) Relationship between grating depth and laser efficiency. Redrawn from [32]

PMMA is a low-cost and environmentally stable material. In literature, there were different sample designs tested for dye-doped PS[14, 29–32]. However, the research using PMMA as a thin-film laser was not fully developed. Literature regarding the direct integrating of grating structure on PMMA film could barely be found. In this work, we would like to investigate the usage of dye-doped PMMA as an organic dye laser. Therefore, different sample designs were proposed. The laser characteristics of the sample designs as polymer thin-film lasers were evaluated to find out the sample model with the best performance. In the best sample model, the layer thickness of the gain material and the resonator were modified to assure a good outcome. Besides, the dye concentration of the sample was changed to understand its influences on lasing properties. Moreover, the effect of different grating periods on the lasing properties was studied. Furthermore, the feasibility of other dyes with PMMA as organic dye laser was tested. In addition, an application of the organic dye-doped PMMA thin-film laser was investigated.

4 Experimental Methods

In this chapter, a short description of the used materials (the matrix material, the photoresist, and the dyes) is illustrated. Besides, the fabrication methods of samples are displayed. Here, there are in total six sample designs proposed to find out the sample model with the best performance. Furthermore, a variety of methods are applied to measure the sample characteristics. In general, the characteristics are categorised into three divisions: sample outlook, material characteristics, and lasing properties. The measurement methods are briefly described in this chapter.

4.1 Material

4.1.1 Poly(methyl methacrylate) (PMMA)

PMMA served as a matrix material for the organic dye in this research. The material is a transparent thermoplastic material that is often used as an alternative to glass. The role of PMMA is nowadays very important for our daily life. One of the main inventors of this material, Dr. Otto Röhm, found a method to fabricate this material commercially in the early nineties [123]. The PMMA used in this work, PLEXIGLAS®8N, was purchased from his company, Röhm GmbH, Darmstadt, Germany. The material was supplied in a uniform size as pellets, which were handled on injection moulding machines with 3-zone comprehensive purpose screws. According to the given datasheet, this product has a high mechanical strength by a tensile modulus of 3300 MPa according to the standard of ISO 527. Based on the standard of ISO 11357, the glass transition temperature is 117 °C. Under a temperature 22.8 °C, the refractive index of PMMA is 1.4926 at the wavelength of 589 nm. As shown in Figure 4.1, the absorption spectrum of PLEXIGLAS®8N shows that this material does not absorb any light in the range of visible light. Based on this property, a Nd-YAG laser with second harmonic generation at the wavelength of 532 nm was selected as the excitation source in this work. Appropriate dyes, which can be excited by a 532 nm laser, were chosen to dope this material.

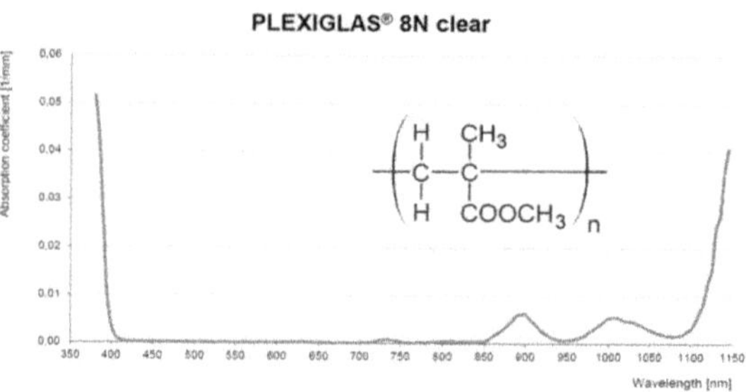

Figure 4.1: Chemical structure of PMMA and its absorption spectrum. Provided by Röhm GmbH.

4.1.2 EpoClad

Another material used in this research is EpoClad, which functioned either as an encapsulation for the samples or as a medium for the grating structure. It was purchased from micro resist technology GmbH, Berlin, Germany. This material is a chemically amplified negative-tone photoresist, which is specially constructed for optical applications. On top of this, the attenuation is low, which is only about 0.2 $dB\,cm^{-1}$ at 850 nm. At 25 °C, the refractive index is 1.54. Together with another product (EpoCore) as a core material, EpoClad is typically used as waveguides for research purposes [124, 125]. It has excellent thermal stability and high transparency to visible light. The supplied EpoClad was a very viscous solution, but it could be diluted by using gamma-Butyrolactone (Gbl) in this research. Figure 4.2 shows that the material is UV sensitive (wavelength < 350 nm), so UV light can be exploited to cure the material while the curing of EpoClad can be performed under a normal air atmosphere.

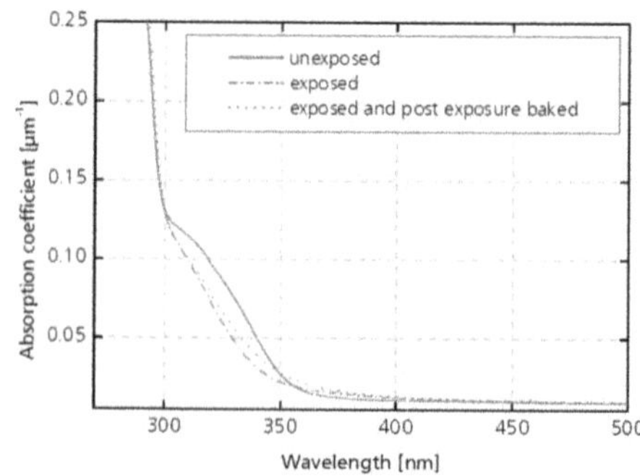

Figure 4.2: Absorption spectrum of EpoClad. Provided by micro resist technology GmbH.

4.1.3 Dyes

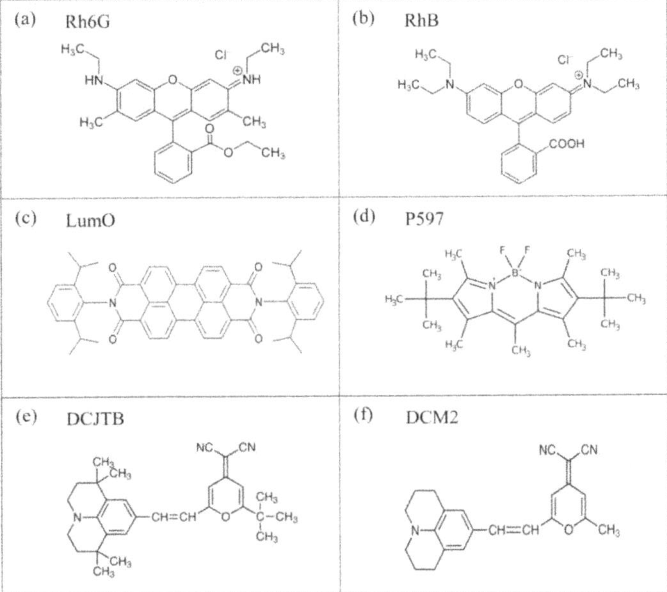

Figure 4.3: Six different laser dyes used in this research work.

As stated in Chapter 4.1.1, PMMA allows transmission in the range of visible light. By taking advantage of this material property, a commercial green light ND-YAG laser with second harmonic generation at the wavelength of 532 nm was used as an excitation source. Therefore, laser dyes that can be excited at 532 nm should be taken into accounts as the dopants of PMMA in this work. By looking into the literature and by following the recommendations from different suppliers, several laser dyes were used in a trial run. The dyes, which showed good lasing characteristics, were Rh6G, RhB, LumO, P597, DCJTB, and DCM2, while their chemical structures are displayed in Figure 4.3.

4.2 Fabrication Methods

Spin-coating

Spin-coating is commonly used in the academy and industry for the deposition of thin-films with high precision [126]. Since the 1990s, this type of solution-processed mechanism has been utilised in the fabrication of organic devices [127, 128]. By using this technique, the samples used in this work were produced.

Figure 4.4 illustrates the concept of spin-coating, which relies on the spinning of a wetted substrate's surface. At first, a substrate is fixed onto a chuck by using a vacuum. Next, the solution is injected on the substrate. Then, the substrate is spin-coated at a constant speed. Due to the angular acceleration from stationary to a specific rotation speed (unit: rounds per minute (rpm)), centrifugal force is formed which induces radial liquid flow. During the progress of angular acceleration, the solution spreads out. Eventually, a uniform film with homogeneous layer thickness can be achieved by using a high and constant rotational speed for a particular time. The resulting thickness d_{sp} can be predicted by using the equation below:

$$d_{sp} = (\frac{\eta}{4\pi\rho\omega_{sp}^2})^{\frac{1}{2}} t_{sp}^{\frac{-1}{2}} \tag{4.1}$$

whereas η is the viscosity coefficient of the solution, ρ is the solution density, ω_{sp} is the angular velocity, and t_{sp} is the spinning time. Solvent evaporation and non-Newtonian behaviour of the solution [129, 130] were not taken into account in this work.

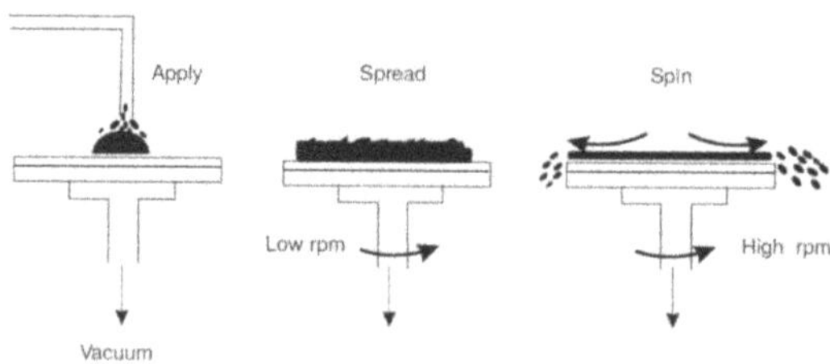

Figure 4.4: Spin-coating processes. Reprinted from [131].

Simulation of Refractive Index Profile

The thickness of the gain material is essential to assure that only the fundamental mode propagates. As mentioned before, the simulation program from [44] was applied to calculate the thickness needed. Meanwhile, the fill factor resulting from this thickness could also be determined.

Several dyes were used to dope with PMMA, in which the light emission was at the range from 550 nm to 600 nm, approximately. Therefore, the simulation was done at the wavelength of 575 nm. For this simulation, the oxide film of the silicon substrate with a thickness of 1500 nm was also taken into account. Besides, a 1000 nm layer thickness of air was also considered in the simulation. Figure 4.5 shows the refractive index profile of the layered system and the profile of the TE_0 modal intensity at the wavelength of 575 nm. According to the simulation, the optimal thickness of the sample produced denoted 1200 nm. Furthermore, the fill factor Γ_{PMMA} was determined as 91.6%.

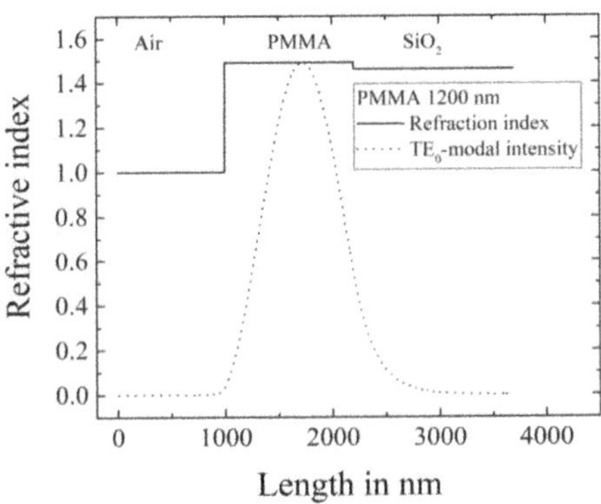

Figure 4.5: Refractive index profile and TE_0 modal profile of the dye-doped PMMA according to the simulation in [44].

Stamp Production

In this work, different designs of DFB lasers were taken into consideration to find out the sample model with the best result. Therefore, it was essential to fabricate DFB thin-film samples with appropriate grating periods. In order to produce the samples, the technique, soft lithography, was used to produce the stamp with appropriate grating according to Equation 3.13 [104, 132]. The material used for the stamp production was polydimethylsiloxane (PDMS), SYLGARD™184 Silicone Elastomer, which was purchased from MAVOM GmbH, Steinfurt, Germany. PDMS is highly transparent which allows easy inspection of components. By mixing the base and the curing agent in a proper ratio, the curing process could be easily controlled by temperature. The operating temperature of this product is at a range from −45 to 200 °C.

The PDMS stamp used in this research was achieved by reproducing the grating from the master stamp used in [44]. The base and the curing agent of PDMS were mixed in a ratio of 10:1 in a beaker. The mixture was stirred until it was cloudy to ensure that the base and the curing agent were completely mixed together. Afterwards, the mixture was vacuum-sealed in the sluice of a glovebox to remove the gas bubble in the mixture. Meanwhile, the master stamp and a petri dish were cleaned by using acetone. Next, the master stamp was laid on the petri dish with the grating structure pointing upwards. Then, the mixture was carefully poured over the master stamp in the petri dish. It was important that no gas bubble was formed. At the meantime, the oven was pre-heated at 75 °C for 15 min. Later, the mixture was heat-cured for 75 min. Eventually, PDMS was cured and was cooled down to room temperature. Then, it was peeled-off from the master stamp. The PDMS stamp with the required grating was cut by using a sharp cutter. Figure 4.6 shows a PDMS stamp with six grating periods in the range from 370 nm to 420 nm with a 10 nm difference to the next grating period.

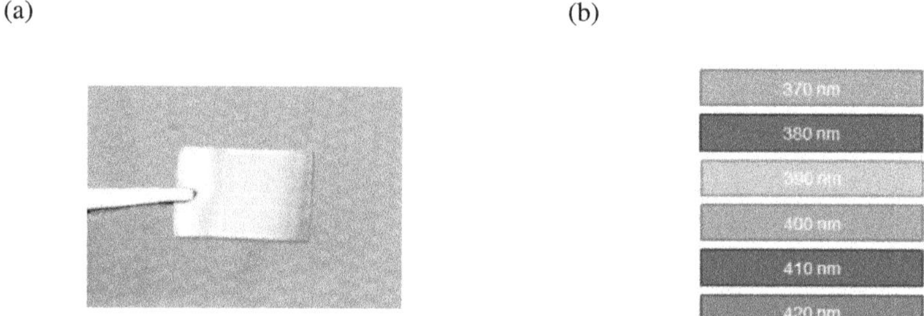

Figure 4.6: (a) A PDMS stamp with grating periods from 370 nm to 420 nm.

Dye concentration

The unit used for the dye concentration in PMMA was parts per million (ppm). It was defined by comparing the ratio of the mole of dye (n_{dye}) to the mole of PMMA (n_{PMMA}):

$$dyeconcentration = (\frac{n_{dye}}{n_{PMMA}}) * 1e - 6 \tag{4.2}$$

Meanwhile, n_{dye} and n_{PMMA} could be calculated by using the following equations:

$$n_{dye} = (\frac{W_{dye}}{MW_{dye}}) \tag{4.3}$$

$$n_{PMMA} = (\frac{W_{PMMA}}{MW_{MMA}}) \tag{4.4}$$

whereas, W_{dye} is the weight of the dye, MW_{dye} is the molecular weight of the dye, W_{PMMA} was the weight of the PMMA and MW_{MMA} was the molecular weight of the monomer of PMMA ($100.12\,\mathrm{g\,mol^{-1}}$).

Sample Fabrication

To find out what kind of sample design was the best structure as an organic dye laser, six sample structures were suggested. A few solvents were used to dissolve the matrix material (PMMA) for sample production, while butanone demonstrated the best result (see Chapter 5.1). For design comparison, the concentration of PMMA in butanone was kept at 7.5wt%, so that the distribution of dye in PMMA was constant. Meanwhile, Rh6G was selected as the dye dopant for PMMA, because Rh6G was commonly doped in PMMA for the research of fiber laser in [133–135]. The concentration of Rh6G was kept at 400 ppm for comparison. Together with the Rh6G, the PMMA pellets were dissolved in butanone to achieve a mixture of gain material. EpoClad, which functioned either as an encapsulation layer or as a medium for grating structure, was diluted with Gbl so that the EpoClad was 50wt%. All the dye-doped PMMA was firstly spin-coated at 350 rpm. If EpoClad was used, it was spin-coated at 4000 rpm. The silicon (Si)-substrate used here had a dimension of $20 \times 20\,\mathrm{mm}^2$ with a silicon dioxide thickness of 1500 nm to avoid absorption of substrate mode.

The first design (D1) of this experiment was the most typical design, which consisted of merely a thin-film formed on top of the substrate. PMMA pellets were dissolved together with Rh6G in butanone. The mixture was directly spin-coated on a substrate to achieve the film structure (see Figure 4.7). In addition to the lasing properties, the material characteristics of different dyes in PMMA were measured by using the D1 structure.

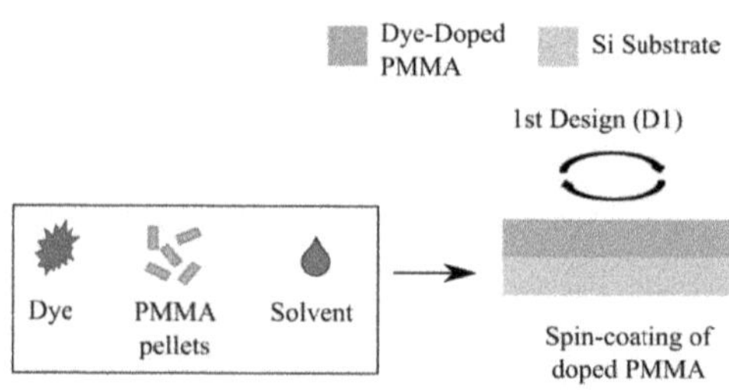

Figure 4.7: Fabrication procedure of the first design (D1).

In the second design (D2), a grating structure was introduced on the gain material (see Figure 4.8). The appropriate grating period was chosen according to Equation 3.13. At first, the gain material was spin-coated on the substrate to obtain a thin-film. Next, a PDMS stamp with the suitable grating period was placed on top of the thin film. Together with the PDMS stamp, the sample was heated at 130 °C and placed under a pressure of 6.125 mN mm^{-2} for 25 min, simultaneously. Afterwards, it was cooled down. The PDMS stamp was peeled-off for measurements later on.

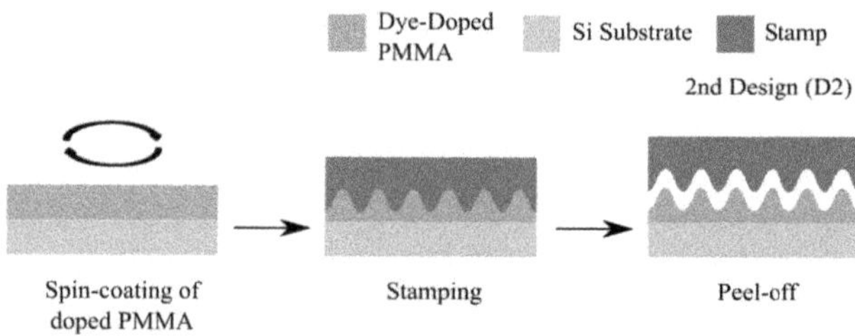

Figure 4.8: Fabrication procedure of the second design (D2).

In the third design (D3), a layer of photoresist, which was EpoClad, was recommended on top of the film structure (see Figure 4.9). First, a film structure was obtained by spin-coating the gain material on top of a substrate. Next, a layer of EpoClad was spin-coated on the gain material film. Then, the sample was pre-baked at 120 °C for 5 min. Afterwards, it was cured under UV light for 20 s. As the last step, the sample was post-baked at 120 °C for 3 min. The sample was cooled down before any experiments.

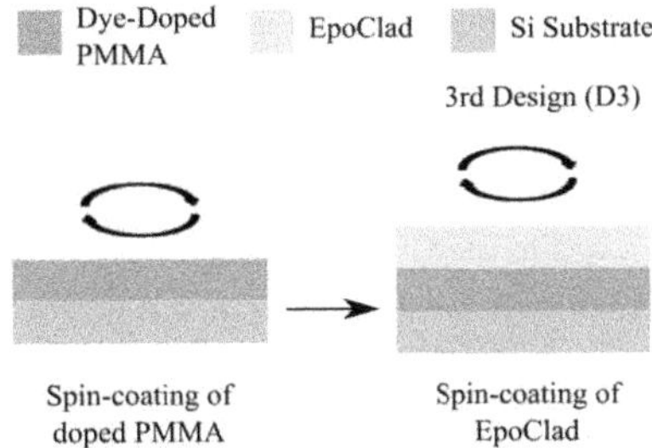

Figure 4.9: Fabrication procedure of the third design (D3).

An EpoClad layer was introduced to the gain material with a grating structure in the fourth design (D4) (see Figure 4.10). First of all, the mixture of gain material was spin-coated on the substrate. Then, a PDMS stamp with an appropriate grating period according to Equation 3.13 was placed on the film of gain material and was heated at 130 °C. At the same time, the pressure was kept at $6.125\,\mathrm{mN\,mm^{-2}}$ for a duration of 25 min. Later, the PDMS stamp was removed. Until here, the process was the same as the production of D2. Then, a layer of EpoClad was spin-coated on the grating structure. Next, the sample was pre-baked at 120 °C for 5 min. After UV-curing for 20 s, post-baking took place at 120 °C for 3 min. Before any experiments, it was cooled down to room temperature.

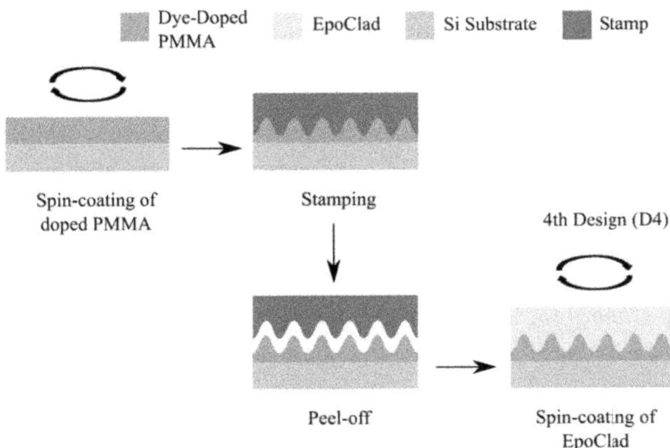

Figure 4.10: Fabrication procedure of the fourth design (D4).

In the fifth design (D5), the grating structure was established on the EpoClad layer instead of the doped PMMA layer (see Figure 4.11). In the first step, a film of grating material was manufactured on top of the substrate through spin-coating. Then, a layer of EpoClad was spin-coated on the film of doped PMMA. The sample was then pre-baked at 120 min for 5 min. Next, the sample was cooled down for about 10 s. Afterwards, a PDMS stamp with the suitable grating period according to Equation 3.13 was located on top of the sample. Together with the PDMS stamp, the sample was cured under UV light for 20 s. Later, it was post-baked at 120 min for 3 min. Eventually, the PDMS stamp was removed and the sample was ready for measurements.

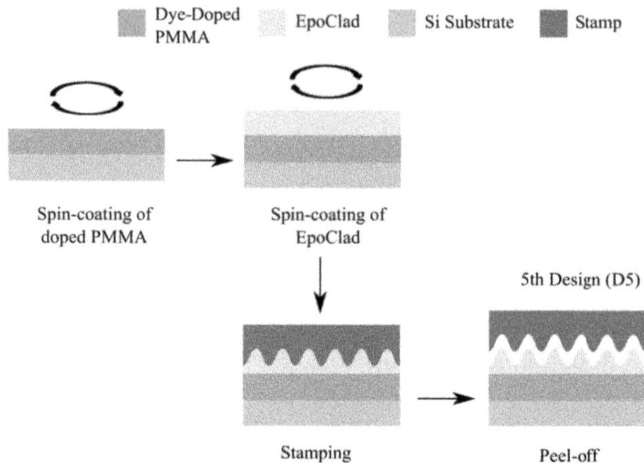

Figure 4.11: Fabrication procedure of the fifth design (D5).

In the sixth design (D6), the EpoClad layer with a grating structure was established on top of the substrate primarily before the doped PMMA was spin-coated on top of it (see Figure 4.12). At first, EpoClad was spin-coated on the substrate. Next, pre-baking of the sample was carried on at 120 °C for 5 min. After cooling down for about 10 s, a PDMS stamp with the grating period based on Equation 3.13 was attached on the EpoClad layer. Then, UV-curing of the sample together with the stamp was performed for 20 s. Later, the sample together with the PDMS stamp was post-baked at 120 °C for 3 min before peeling-off. The sample was cooled down again for about 10 s. Afterwards, doped PMMA was spin-coated on top of it.

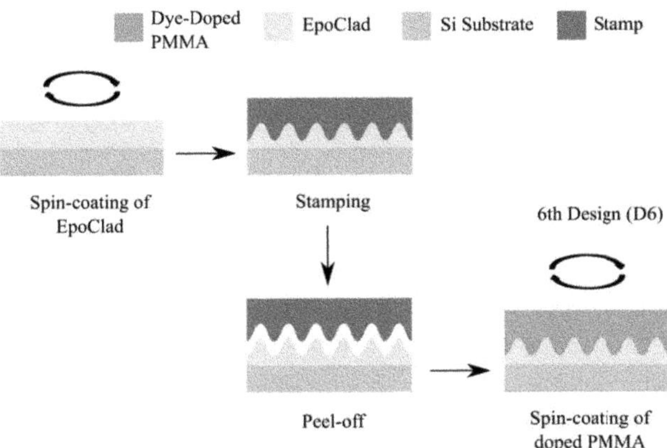

Figure 4.12: Fabrication procedure of the sixth design (D6).

4.3 Measurement Methods

4.3.1 Sample Outlook

Profilometry is an approach to obtain topographical data from a surface of samples, which can be a single point, a line scan or even a full three-dimensional scan. The profilometer, Dektak 8, which was purchased from Vecco was used to measure the thickness of the samples in this work. It was a stylus profilometer, which used a physical probe to detect the surface. The probe moved along the surface in order to acquire the surface height, which was used to define as the thickness of the samples, eventually. To achieve this purpose, scratches were done on the surface in advance. The force from the samples pushing up against the probe during the scans was monitored, so that a specific amount of torque was maintained. The changes in the Z position of the arm holder was used for the determination of the layer thickness [136].

Another method applied was atomic force microscope (AFM), which is a technique of scanning probe microscopy with nanometer resolution. It contained a cantilever with a tip, which was in contact with the samples either in contact mode or contactless mode. The AFM used in this research was CombiScope™100, purchased from Aist-Nt Co., which belonged to Dresden Integrated Center for Applied Physics and Photonic Materials (IAPP), TU Dresden. It was used to measure the surface properties of the samples, which were the roughness of the surface of the sample and the depth of the grating structure.

Furthermore, a scanning electron microscope (SEM) was also used to study the surface topography of the samples. SEM is a kind of electron microscope, which can provide images of a sample by scanning the surface using a focused beam of electrons. The information of surface structure and composition of the sample could be generated by the interaction of the electrons with atoms in the sample. The samples could be sputtered with gold atoms for further studies. The SEM used was EVO LS 25 purchased from Zeiss, which belonged to Institute for Chemical and Thermal Process Technology (ICTV), TU Braunschweig.

4.3.2 Material Characteristics

Absorption and Photoluminescence Spectrum

For the measurement of the absorption spectrum of the samples, the spectrophotometer, Lambda 9, was implemented, which was purchased from Perkin-Elmer, Rodgau, Germany. The instrument had two beams, in which one of the beams was used as the reference for the measurement. The range of the wavelength supplied by the instrument was from 185 nm to 3200 nm. Whereas, only the range of visible light was concerned for this research. Afterwards, the absorption coefficient was calculated by using data from the absorption spectra based on Equation 3.2. For the measurement of the absorption spectrum, the selected dye and PMMA pellets were dissolved in butanone, before spin-coating on a transparent borosilicate substrate with a dimension of $20 \times 20\,\text{mm}^2$. The doped PMMA layer consisted of a film thickness of 1200 nm.

The samples on borosilicate substrate could be further applied in the setup in Figure 4.13 to define the photoluminescence spectrum. The ND:YAG laser, STA-01-1-1SH (purchased from Standa, Vilnius, Lithuania), was applied to excite the sample, in which the pump energy was maintained at 17.69 nJ by using a Glan-Thompson polariser. The pump energy was kept low but the emission still could be observed by the spectrometer, Triax320, which was purchased from Horiba Jobin Yvon, Bensheim, Germany. In such a way, ASE could be avoided.

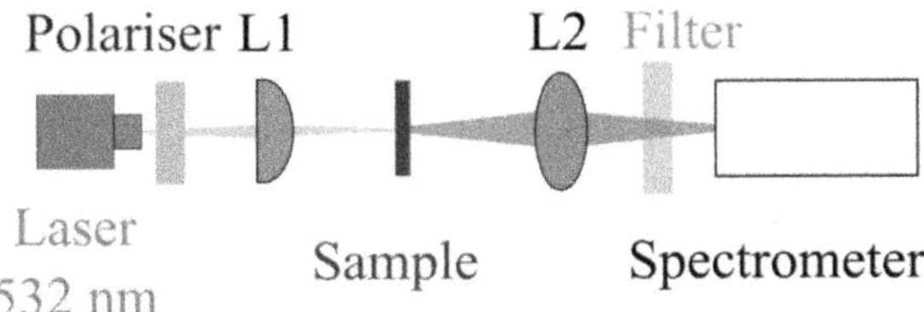

Figure 4.13: Photoluminescence setup.

PLQY and Fluorescence lifetime

The definition of PLQY of material is defined as the ratio of the number of emitted photons to the number of absorbed photons. The product, FluoTime 300 "Easy Tau", purchased from PicoQuant, Berlin, Germany was used for this measurement. An integrating sphere was installed in this instrument, which was designed to obtain a homogenous distribution of optical radiation. Due to the multiple Lambert reflections at the sphere's inner surface, all the emitted and scattered light from the samples could be detected by the detector in the integrating sphere. For the PLQY measurement, the wavelength of the excitation source was set at 532 nm. The dye-doped PMMA was spin-coated on a transparent borosilicate substrate with a dimension of $20 \times 20\,\mathrm{mm}^2$ for this measurement.

Fluorescence lifetime is another important feature of fluorescence to characterise the samples. To define fluorescence lifetime, FluoTime 300 was also implemented, in which the measurement setup was modified as Figure 4.14. The sample holder was customised, so that a film-sample can be applied. The dye-doped PMMA layer was produced on top of a $20 \times 20\,\mathrm{mm}^2$ Si-substrate so that the emission light can be reflected for detection. To measure fluorescence lifetime, the method called Time-Correlated Single Photon Counting was applied. By using this method, the fluorescence lifetime was evaluated by measuring the time between the sample excitation by a pulsed laser and the arrival of the emitted photon at the detector. For the determination of fluorescence lifetime, the wavelength of the pulsed laser was set at 532 nm as an excitation source.

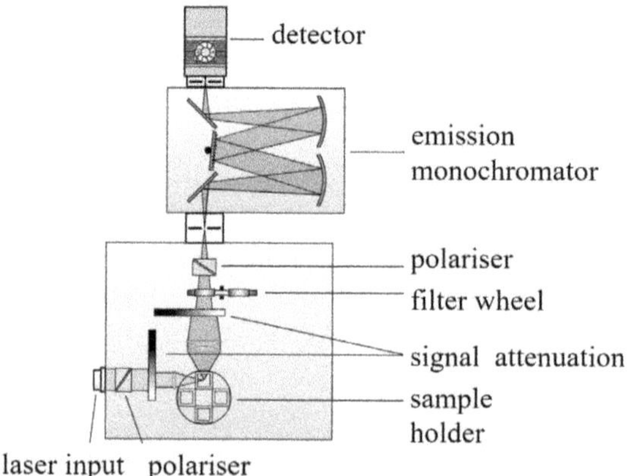

Figure 4.14: Setup in FluoTime 300. Reprinted from the PicoQuant.

Optical Gain

The method used for the determination of optical gain was the variable stripe length (VSL) method. It was proposed by Shaklee et al. in the seventies [137] and is still been practised nowadays [138, 139]. Under this concept, the behaviour of the edge emission of a sample depending on the length of an excitation segment on the sample surface was examined. In this research, thermally oxidized Si substrates with a SiO_2 thickness of 1.5 µm were employed. The Si substrates were prepared in a dimension of $20 \times 20\,mm^2$. When the Si substrates were cut, a smooth edge can be achieved to assure a good out-coupling, which was very essential for detection by using the VSL method. Besides, the oxide layer of the Si substrate was also very important because it prevented the spread of substrate modes which could induce measuring errors in the measurement. By using the VSL method, the optical amplification in a thin-film active device was examined, which was normally referred to as optical gain.

The VSL method is based on the relationship between the excited length and the emitted radiation intensity. By using this relationship, the optical gain can be calculated by applying Equation 3.8. From this equation, three different forms of curves can be generally derived. For high optical gain, there is an exponential increase in intensity. The spontaneously emitted intensity components are added up. If there is no gain, the light intensity increases linearly. If the gain is negative, it is defined as optical absorption or optical loss. These situations are depicted in Figure 4.15. It also shows that both the optical gain and the optical absorption can be determined by the VSL method. The value of the optical gain obtained by this method includes the losses due to scattering or the losses of the evanescent field in the layers. The advantage of VSL is that the structure is very similar to the application of an optically pumped laser.

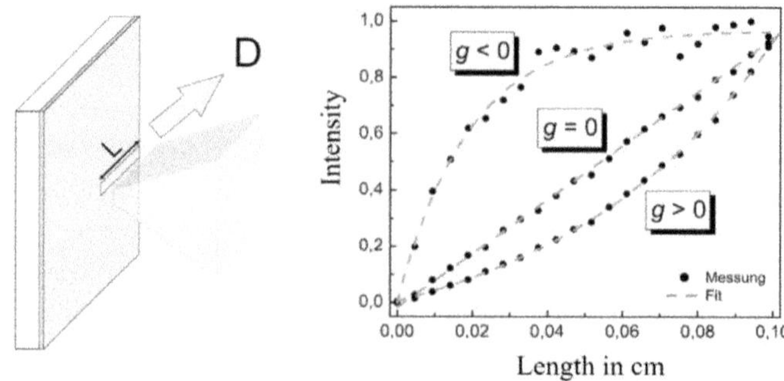

Figure 4.15: By using the VSL method, the intensity is measured at the sample edge by the detector (D) over the length (L) of the excitation area for different signs of gain (g). Reprinted from [44]

Nonetheless, Equation 3.8 is strictly valid for one-dimensional cases, i.e. in a glass fiber. Therefore, there is a two-dimensional problem for a sample with a thin-film structure. The light can propagate unrestrictedly in the film plane, which means in the x and z plane. If the geometry of the VSL arrangement is considered, Equation 3.8 only applies to high amplification, because the intensity components dominate the edge emission. To run the VSL measurement, it is essential to make sure that the gain parameter $g(\lambda)$, the excitation density, and the coupling efficiency of the detector are independent of z. Otherwise, the mathematical model must be adapted to the model in [140]. The requirements can be easily fulfilled by a homogenous structure with low optical gain. With the high optical gain, gain saturation occurs because the intensity increases to an extent that the finite population of the excited states cannot maintain the gain. To avoid such an effect in the experiment, the excited length was reduced during the measurement by the rule of $gL < 4$ depending on the obtained gain [141]. In such a way, the intensity of the mode was scaled down and the saturation effect was prevented.

Figure 4.16 shows the setup of the VSL method. As the source for line-shaped excitation, an ND:YAG laser (STA-01-1-1SH purchased from Standa, Vilnius, Lithuania) was implemented. It delivered excitation pulses up to a repetition frequency of 100 Hz with a pulse duration <600 ps and a single pulse energy up to 500 µJ at the emission wavelength of 532 nm. Several cylindrical lenses and two apertures were applied to achieve the homogenous central part of the expanded Gaussian laser beam (see Figure 4.17). As a result, an around 0.228 mm wide and about 3 mm long excitation line was generated on the samples, while the length of the excitation line still could be varied. A Glan-Thomson polarisation filter was located in front of the laser to control the excitation energy. The edge emission of the sample was focused using a collecting lens on a spectrometer (Triax320 purchased from Jobin Yvon). A Labview®program written in [44] was used for running the measurements. In a measurement, there were approximately 25 measurement cycles/ spectra written into a file, in which only 20 spectra were taken into consideration for the analysis. The file was evaluated with the help of Levenberg-Marquardt algorithm to determine the optical gain.

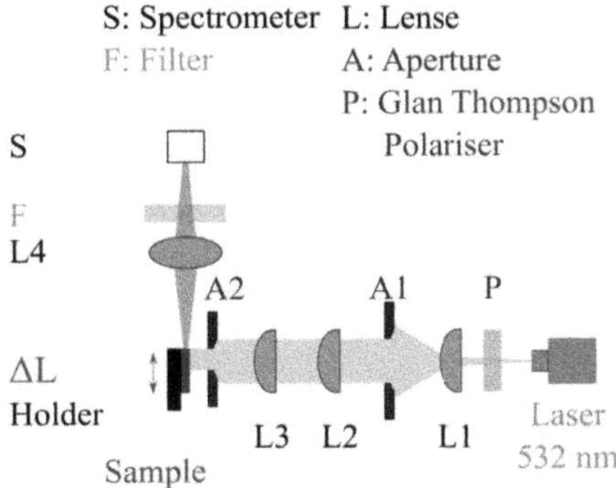

Figure 4.16: Setup of VSL measurement.

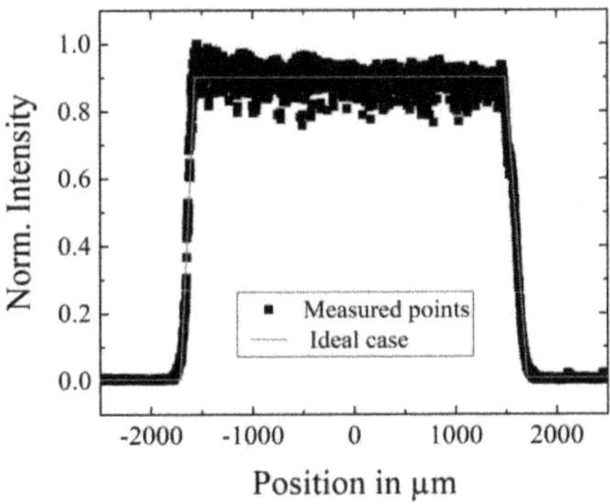

Figure 4.17: Light profile of excitation light of VSL measurement.

4.3.3 Lasing Properties

To distinguish laser from other phenomena like ASE source, there are certain criteria for laser. According to Samuel et. al. [142], a laser should have the following properties: first, the laser consists of a narrow FWHM; second, there is a beam as light output; third, a clear lasing threshold must be given; forth, the light emission can be affected by the specific gain medium and resonator. To claim as a laser, typical results shown in literature are lasing spectrum and lasing threshold, which were also determined in this work. In addition to lasing spectrum and lasing threshold, polarisation extinction ratio (PER), and output energy of the samples were also investigated.

Lasing Spectrum and Lasing Threshold

The setup shown in Figure 4.18 was employed to evaluate the lasing spectrum and the lasing threshold of the sample. This laser setup was nearly identical to the setup used for the VSL method. The sample was excited transversally or perpendicularly to the sample surface so that optical losses due to total internal reflection were avoided. In this measurement, the excitation length of the sample was fixed at 2.3 mm. The homogeneity of the excited line is shown in Figure 4.16 (b). To measure the spectrum, a nitrogen-cooled spectrometer, Triax320 (purchased from Jobin Yvon, Bensheim, Germany) was applied. Several spectra of the sample were taken under various excitation energies, which were evaluated using the software, OriginPro, eventually. For noticing the narrowing of FWHM, two spectra before the lasing threshold (0.10 μJ) and after the lasing threshold (1.69 μJ) were drawn. Besides, the maximum intensity of the spectra with a series of pump energies was noted and normalised to its highest value, so that a curve of the pump energy against the measured maximum intensity could be drawn. The turning point, in which the light intensity was amplified, was defined as the lasing threshold.

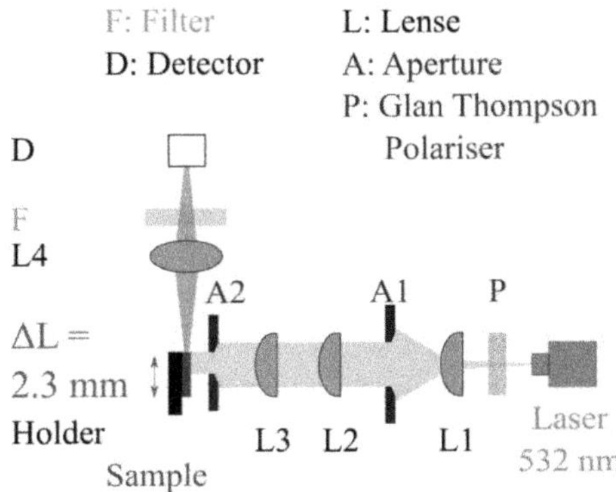

Figure 4.18: Laser setup for the measurement of lasing properties.

Polarisation Extinction Ratio (PER)

PER compares the ratio of the optical power held on the "wanted" axis to that which is on the "unwanted" axis. The value is significant in telecommunications. The "wanted" axis is considered as the principal linear polarisation state (transverse electric (TE)), while the "unwanted" axis denotes the orthogonal polarisation state (transverse magnetic (TM)). To measure PER, the rotating polariser method was implemented. A polariser, LPVISE100-A, (purchased from Thorlabs, Bergkirchen, Germany) was established in front of the detector (s. Fig 4.18). Spectra were taken at 0° and 90°, which represented the spectra at the maximum power and the minimum power, respectively. Meanwhile, the maximum power represented the principal linear polarisation state, while the minimum power corresponded to the orthogonal polarisation state. The maximum value of spectra at 0° was denoted as P_{max}, while the maximum value of spectra at 90° was referred to as P_{min}. With these value, PER could be illustrated in decibles (dB) by using the following equation:

$$PER = 10log_{10}\frac{P_{max}}{P_{min}} \tag{4.5}$$

Output Energy

The setup to measure the output energy of the sample was exactly the same as in Figure 4.18. For the measurement of the pump energy and the output energy from the samples, the power/energy meter, LabMax™-TOP (purchased from Coherent, California, US) was applied. To have correct measurements, the measured wavelength of the power/energy meter should be adjusted in advance. The frequency of the measured beam could be directly read out from this instrument. Proper power/energy sensors needed to be selected to evaluate the output energy. In this experiment, power/energy sensors, J-10SI-LE (measured range from 20 pJ to 40 nJ) and J-10MT-10KHZ, (measured range from 100 nJ to 200 μJ) were used.

Other Features

On top of the characteristics mentioned, other features like sample lifetime and pulse duration were also investigated. To determine the sample lifetime, pump pulses was used as the unit. The sample lifetime was determined in a way that how many pump pulses could the sample sustain, in which the maximum intensity dropped to its 50%. For the aspect of pulse duration, a signal analyser, R&S FSV, (purchased from Rhode & Schwarz GmbH, Munich, Germany) together with a free-space Si Detector, DET025/M, (purchased from Thorlabs GmbH, Bergkirchen, Germany) were implemented. The FWHM of the pulse, which was emitted from the sample, was evaluated as the pulse duration of the sample.

5 Result

5.1 Sample Outlook

Solvent used

As mentioned before, the PMMA was supplied in the form of pellets. In order to produce thin-film samples, PMMA pellets needed to be dissolved by an appropriate solvent. The used solvent should provide a smooth surface for the samples, eventually. A smooth surface was required for lasing experiments because a rough surface could cause unnecessary scattering. Different solvents were applied to dissolve the pellets of PMMA, which were butanone, chloroform and dichloromethane. Afterwards, the solution was spin-coated into thin-film at 4000 rpm spin-coating speed. A closer look at the surface of the thin-films was taken by using SEM (see Figure 5.1). The surface of the samples using butanone and chloroform were considered to be smooth. On the other hand, there were many holes that could be clearly seen on the films using dichloromethane. This is due to fact that dichloromethane has a relatively lower boiling point, which is only 39.6 °C. Eventually, butanone was selected for a further investigation because butanone has a higher boiling point (79.6 °C) than chloroform (61.2 °C), so that unnecessary evaporation problems could be avoided.

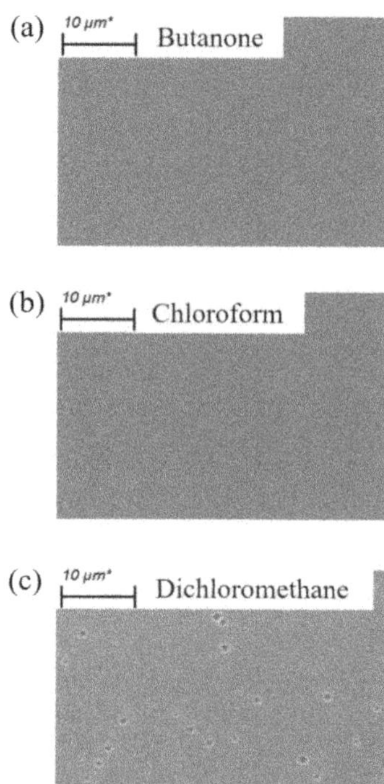

Figure 5.1: SEM images of the sample surface, in which the PMMA pellets were dissolved by using (a) butanone, (b) chloroform, and (c) dichloromethane.

Surface Roughness

A rough surface will induce unnecessary scatterings, which causes inefficiency of laser generation. Therefore, it is crucial to evaluate the surface roughness of the samples to ensure a smooth surface. In this research, there were two different material layers, which were the PMMA layer and the EpoClad layer. The PMMA pellets were dissolved by using butanone and EpoClad was diluted by Gbl prior to spin-coating to achieve a thin-film structure. Afterwards, the surface roughness of PMMA and EpoClad was measured using the technique of AFM (see Figure 5.2). The root mean square roughness of both PMMA and EpoClad was smaller than 1 nm. This result demonstrated that the sample was smooth enough for the experiments later on.

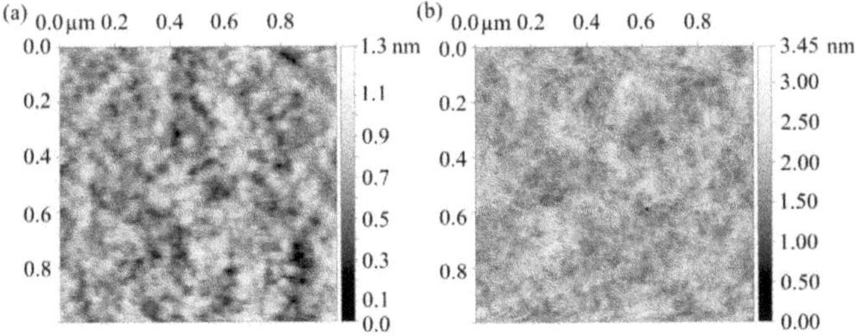

Figure 5.2: AFM images of both (a) the PMMA layer and (b) the EpoClad layer.

Grating Structure

The grating structure was applied in the tested sample designs. In D2 and D4, a grating structure was proposed on the PMMA film. Meanwhile, there was a grating structure on the EpoClad layer in D5 and D6. Therefore, it is essential to check if the grating structure could be successfully built on the PMMA layer and the EpoClad layer. A quick look at the surface profile of the PMMA layer and the EpoClad layer with a grating structure was measured by using SEM (see Figure 5.3 and Figure 5.4). The top view and the side view of the grating structure on the PMMA layer and the EpoClad layer were observed, respectively. It was confirmed that a grating structure could be formed on top of the layer of PMMA and EpoClad by using the suggested fabrication method.

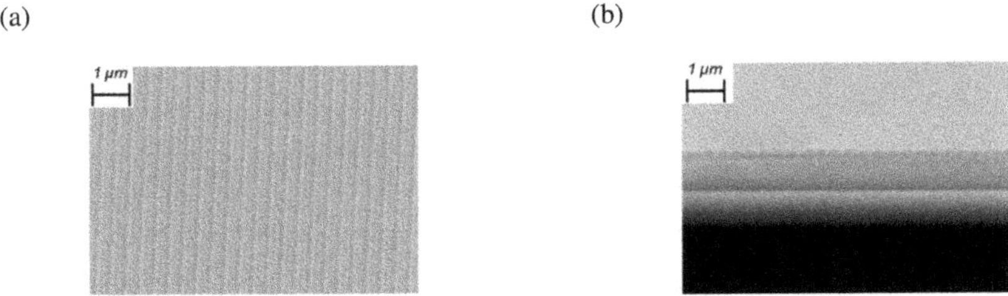

Figure 5.3: SEM image of (a) the top view and (b) the side view of the grating structure on top of the PMMA layer.

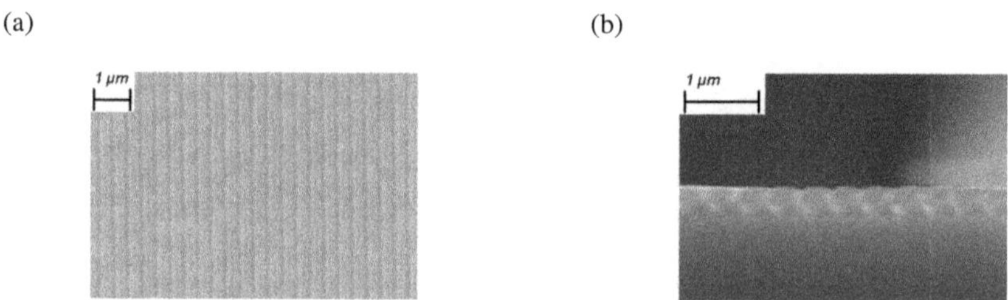

Figure 5.4: SEM image of (a) the top view and (b) the side view of the grating structure on top of the EpoClad layer.

To further understand if the grating structure was properly established onto the PMMA layer and the EpoClad layer, the applied grating period was evaluated by using AFM. The grating period of both PMMA layer and EpoClad layer was intended to be 380 nm. The measured average grating period on top of the PMMA layer was 379.85 nm (see Figure 5.5), while it was 380.16 nm on top of the EpoClad layer (see Figure 5.6). The result was extremely close to the desired value. Therefore, it can be deduced that the grating structure with the wanted grating period can be properly built on top of the PMMA layer and the EpoClad layer. In addition, the grating depth was also measured. The grating depth of PMMA was only around 60 nm, which was smaller than that of EpoClad, which was approximately 80 nm.

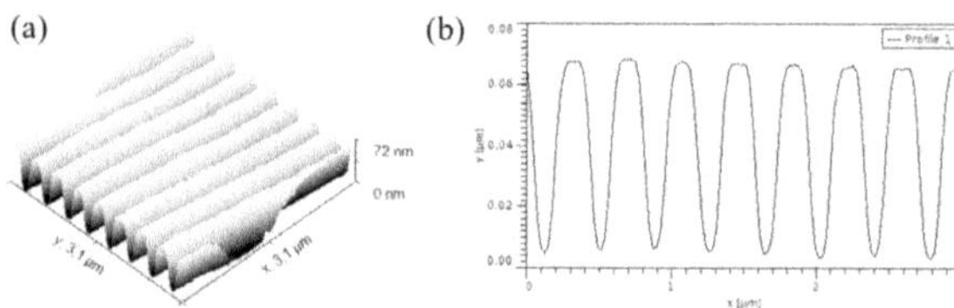

Figure 5.5: (a) 3D AFM image and (b) grating profile of the PMMA layer.

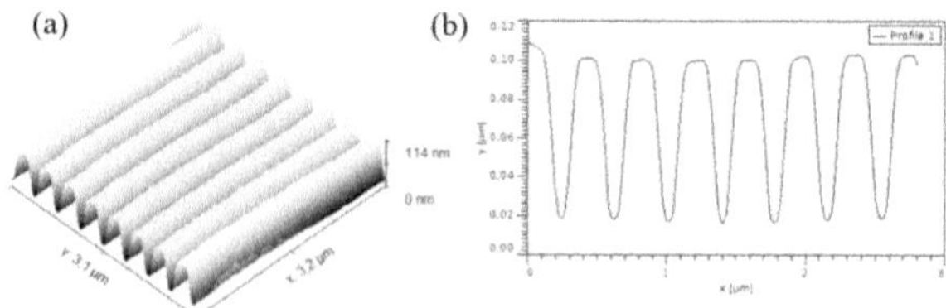

Figure 5.6: (a) 3D AFM image and (b) grating profile of the EpoClad layer.

5.2 Material Characteristics of Samples Doped with Different Concentrations

Rh6G is one of the most common laser dyes found in literature, which is regularly used to doped into PMMA for fiber laser production [133–135, 143–145]. However, the research as an organic dye-doped PMMA thin-film laser was not fully studied yet. Therefore, Rh6G was selected as the first candidate for the studies. In the beginning, different concentrations were used for the test, which were 50 ppm, 100 ppm, 200 ppm, 400 ppm, and 600 ppm. However, for the samples with a concentration of 50 ppm, there was no emission observed. On the other hand, Rh6G with a concentration of 600 ppm could not be fully dissolved because the solution was too saturated. In consequence, only the samples with the concentrations of 100 ppm, 200 ppm and 400 ppm were further examined.

Figure 5.7 shows the absorption spectra and the photoluminescence (Pl.) spectra of the Rh6G with different concentrations. As seen, the maximum value of the absorption coefficient (α_{abs}) became higher with increasing concentration. Eventually, the result of absorption coefficient (α_{abs}) at 532 nm is presented in Table 5.1. In addition, it could also be observed that the peak of the photoluminescence spectra was slightly red-shifted when the concentration got higher. By using 100 ppm Rh6G concentration, the peak wavelength was detected at 561.89 nm. For the concentration of 200 ppm and 400 ppm, the peak wavelengths were measured to be 562.66 nm and 564.74 nm, respectively. Samples doped with higher concentrations indicated a larger number of Rh6G dye molecules in the gain medium. This means that the possibility of reabsorption by a Rh6G dye molecule is much higher. The red-shift was caused by the re-emission of the reabsorbed photon. It was discovered that the concentration of Rh6G did not affect the refractive index of PMMA. Meanwhile, the value of the peak wavelength was used to calculate the appropriate grating period needed for the grating structure based on Equation 3.13, which was 380 nm.

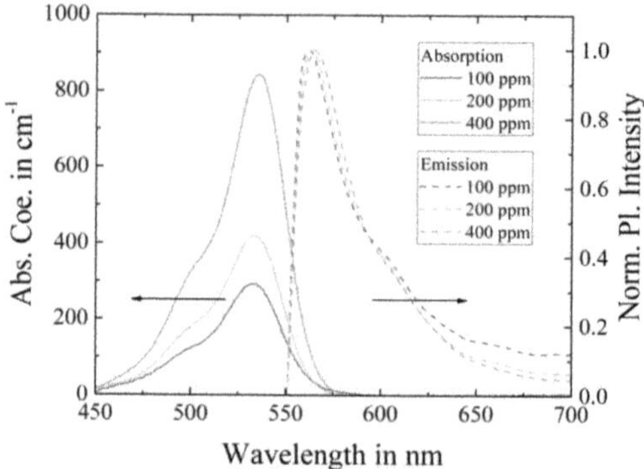

Figure 5.7: Absorption spectra and photoluminescence spectra of thin-film sample with different Rh6G concentrations: 100 ppm, 200 ppm and 400 ppm.

The overlapping area between the absorption spectra and the photoluminescence spectra represented the reabsorption of the dye, which caused inefficiency of emission. The overlapping area was calculated and is shown in Table 5.1, which was normalised to the area of the samples with the concentration of 100 ppm for a better comparison. As the concentration increased, more aggregates formed and the area of overlapping became bigger. The FWHM of the spectra was around 36 nm, which was quite broad because there was only spontaneous emission.

Table 5.1: Absorption coefficient of the Rh6G with different concentrations at 532 nm, FWHM of photoluminescence spectra and normalised overlapping area of the absorption spectra and the photoluminescence spectra.

Rh6G concentration /ppm	α_{abs} at 532 nm /cm^{-1}	FWHM of photoluminescence spectra /nm	Overlapping area compared to 100 ppm Rh6G
100	289.26	35.59	1
200	409.29	36.00	1.14
400	794.40	36.32	2.40

PLQY is another important property to measure since it can give information about the ratio of the absorbed photons to emitted photons. In this study, the PLQY was relatively constant under different concentrations (see Figure 5.8(a)) The value was around 62%. On top of this, the fluorescence lifetime of the sample with different concentrations of Rh6G was tested (see Figure 5.8(b)). For the concentration of 100 ppm and 200 ppm, the fluorescence lifetime was 3.43 ns, while the fluorescence lifetime of the concentration of 400 ppm was 3.44 ns, which meant that the fluorescence time was quite identical under various concentrations. Here, it can be deduced that Rh6G concentration does not affect the PLQY and the fluorescence lifetime.

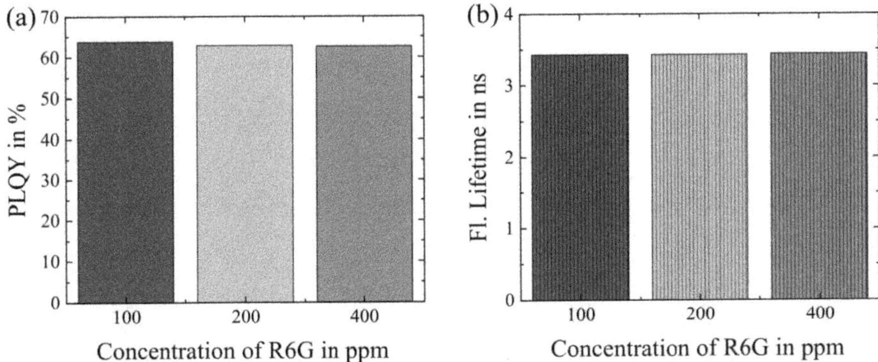

Figure 5.8: (a) PLQY and (b) fluorescence lifetime of thin-film sample with different concentrations: 100 ppm, 200 ppm and 400 ppm.

By using the VSL method, the optical gain of the samples was studied. The value was essential to determine the stimulated emission cross section, which indicates the emission efficiency. The optical gain of the Rh6G thin-film samples with different concentrations: 100 ppm, 200 ppm, and 400 ppm is shown in Figure 5.9. In this measurement, the pump energy density was fixed at $122.40\,\mu J/cm^2$. It was apparent that the maximum value of optical gain was enhanced with increasing concentration. On top of this, the maximum value of optical gain was red-shifted to a longer wavelength because reabsorption happened more frequently at higher dye concentrations. The reabsorbed photons re-emitted, which caused the red-shift. Meanwhile, the maximum value of optical gain of the sample with 400 ppm Rh6G was the highest among the samples as there were more dye molecules with higher concentrations for the contribution of modal gain.

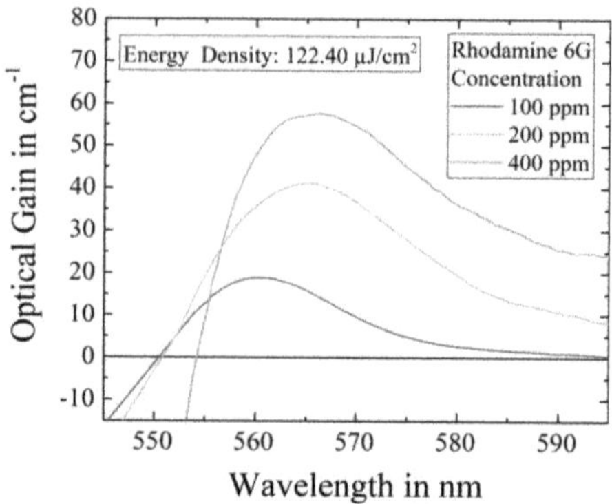

Figure 5.9: Optical gain of the sample doped with different Rh6G concentrations.

In Figure 5.10, the concentration of Rh6G was fixed at 400 ppm, while the optical gain shown was induced by different energy densities: 30.08 $\mu J/cm^2$, 109.44 $\mu J/cm^2$, and 232.72 $\mu J/cm^2$. When the energy density increased, the maximum value of the modal rose. At the same time, it shifted to a shorter wavelength. This is due to the interaction of the Rh6G dye molecules with the polar parts of PMMA. The long hydrocarbon chain and CH_3 are non-polar. On the other hand, $COOCH_3$ groups of PMMA are polar. By increasing the pump energy, the interaction of the Rh6G dye molecules with the polar molecular environment became more intense. There was a blue-shift when the environment was more polar according to [146]. The same condition was also detected in [147].

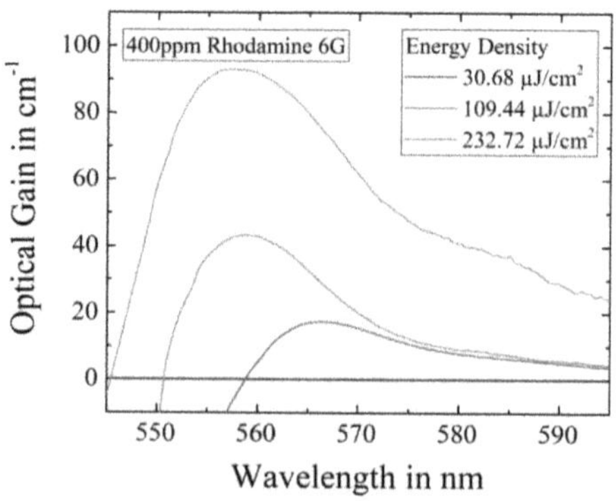

Figure 5.10: Optical gain under the effect of different pump energy densities.

To further understand the trend of optical gain, the samples with different concentrations were excited with a series of energy densities. Figure 5.11 (a) shows the relationship of the maximum value of optical gain versus the excited energy density. When the excited energy density was not too high, the maximum value of optical gain presented a linear dependency on the pump energy density. The slope became steeper with increasing concentration. Upon certain energy density, saturation was observed. The energy density of 122.40 μJ/cm^2 was selected for the calculation of stimulated emission cross section because the maximum optical gain values of all the Rh6G concentrations were still linear to the energy density at this particular excited energy density. On top of this, there was no saturation observed. At 122.40 μJ/cm^2, the maximum optical gain value of 100 ppm was only 17.56 cm^{-1}. It was 45.65 cm^{-1} for 200 ppm. For the concentration of 400 ppm, the maximum value of optical gain was 57.13 cm^{-1}, which was the highest. According to Equation 3.4, these values were used for the calculation of stimulated emission cross section. The result is shown in Figure 5.11 (b). As seen, the higher the concentration, the bigger the maximum stimulated cross section. The maximum stimulated cross section of 400 ppm was the highest, which was 3.68×10^{-17} μJ/cm^2. It was followed by 200 ppm with a maximum stimulated cross section of 2.52×10^{-17} μJ/cm^2. 100 ppm had the lowest maximum stimulated emission cross section among the samples, which was only 1.14×10^{-17} μJ/cm^2. Since the value of 400 ppm was the highest, the dye concentration was fixed at 400 ppm for further experiments. On top of this, the excited energy density was set at 122.40 μJ/cm^2 for further optical gain measurement with other dyes.

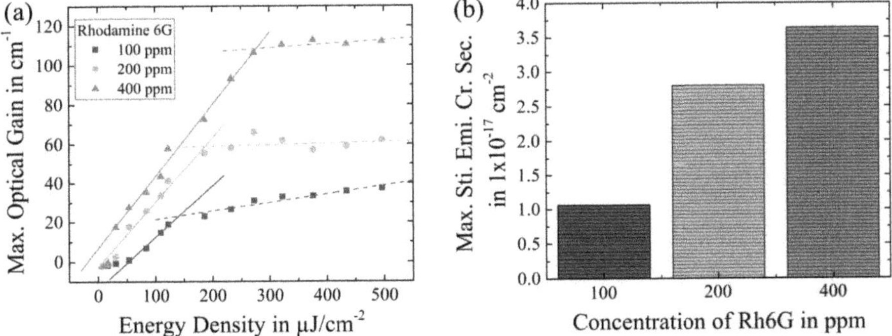

Figure 5.11: (a) Trend of optical gain of the sample doped with different Rh6G concentrations under excitation of a series of energy densities. (b) Maximum stimulated emission cross section of the sample doped with different Rh6G concentrations at 122.40 μJ/cm^2.

5.3 Material Characteristics of Samples Doped with Different Dyes

As mentioned in Chapter 4.1.3, there were in total six different dyes used in this research: Rh6G, RhB, LumO, P597, DCJTB, andDCM2. Each dye was dissolved in butanone together with PMMA pellets before spin-coating for film production. Afterwards, the absorption spectra and the photoluminescence (Pl.) spectra of the film samples were taken (see Figure 5.12 and Figure 5.13). It was important to measure the absorption spectra of the samples to learn if the samples could be excited by a 532 nm laser. By having absorption spectra, the absorption coefficient (α_{abs}) of the samples was calculated using Equation 3.2 and was also shown in Table 5.2. On top of this, the photoluminescence spectra were also essential to provide necessary information about the emission peak wavelength of the samples and the FWHM under spontaneous emission. The peak wavelength of each photoluminescence spectrum was noted in Table 5.2. It was detected that there is no effect from the dye on the refractive index of PMMA and the needed grating periods were calculated by using Equation 3.13.

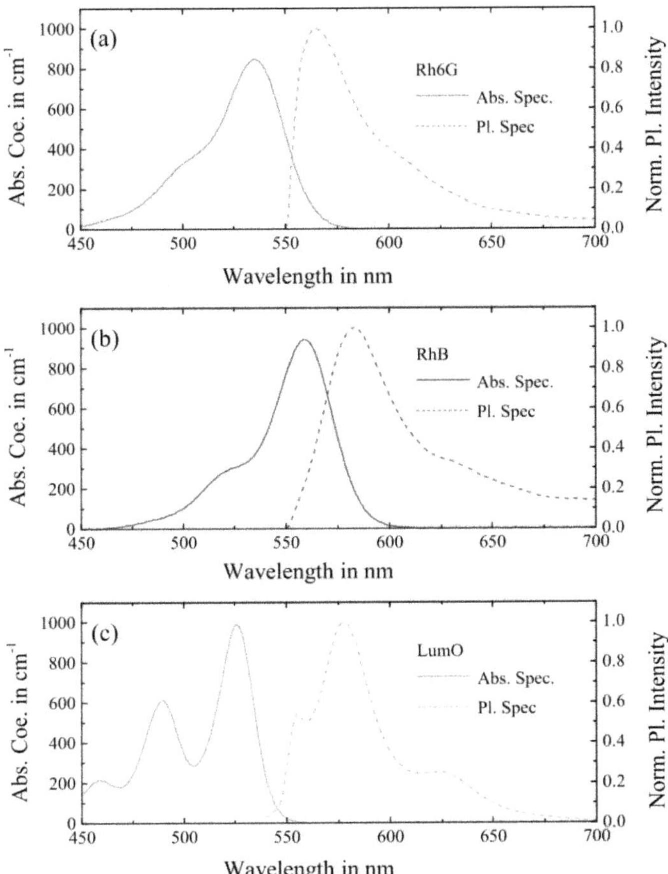

Figure 5.12: Absorption spectra and photoluminescence spectra of the samples doped with different dyes were shown: (a) Rh6G, (b) RhB, and (c) LumO.

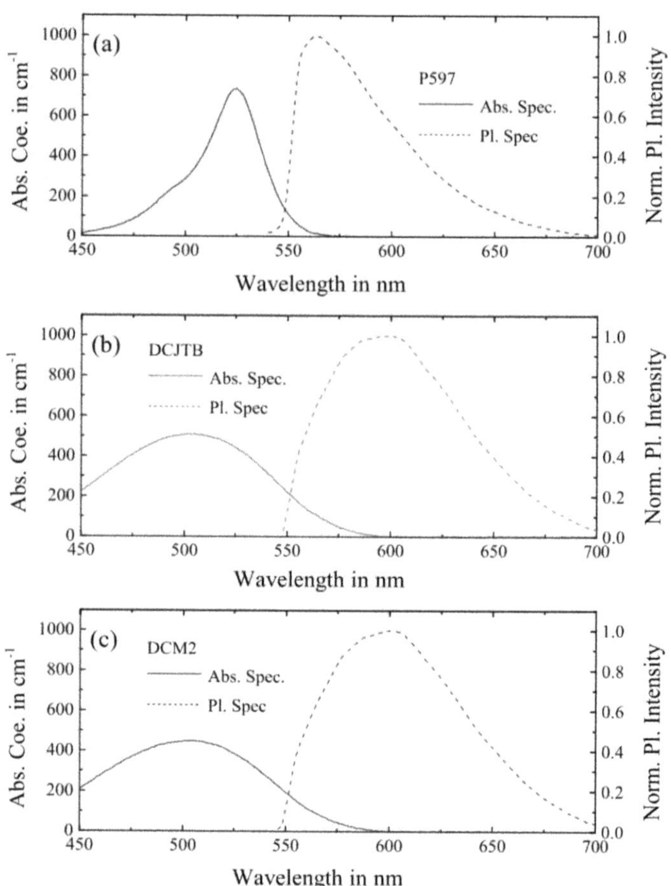

Figure 5.13: Absorption spectra and photoluminescence spectra of the samples doped with different dyes were shown: (a) P597 (b) DCJTB, and (c) DCM2

The overlapping area between the absorption spectra and the photoluminescence spectra in Figure 5.12 and Figure 5.13 indicated reabsorption of the emission, which also demonstrated an inefficiency of the photoluminescence. The result is shown in Table 5.2, which was normalised to Rh6G, as Rh6G was used as a reference. Among the samples, the overlapping area of RhB was the largest. This is due to the fact that RhB has COOH bond, which is not present in other dye molecules. It might indicate that COOH bond caused the big overlapping area of RhB.

Table 5.2: Absorption coefficient of the samples at 532 nm, FWHM of photoluminescence spectra, normalised overlapping area of the absorption spectra and photoluminescence spectra, peak wavelength and grating period.

Dye	α_{abs} at 532 nm /cm^{-1}	FWHM of photoluminescence spectra /nm	Overlapping area compared to Rh6G	Peak wavelength /nm	Grating period /nm
Rh6G	825.04	36.32	1	565	380
RhB	340.07	42.13	3.63	582	390
LumO	739.05	40.64	0.17	577	390
P597	592.97	53.74	0.41	564	380
DCJTB	392.20	86.60	1.09	600	400
DCM2	340.66	83.60	0.98	601	400

The PLQY of the film samples doped with different dyes was determined to investigate the efficiency of the conversion of absorbed photons into emitted photons. The result is shown in Figure 5.14 (a). In general, all the samples had over 60% of PLQY. Meantime, LumO had the highest PLQY among the samples, which was 82.91%. DCJTB had a PLQY value of 76.61%, which was the second-highest. RhB had the third-highest PLQY (73.77%). Next, the PLQY of P597 was 67.79%. DCM2 and Rh6G had quite identical PLQY, which were 62.62% and 62.60%, accordingly. The value of PLQY was also another essential data for the calculation of the stimulation emission cross section, eventually.

In Figure 5.14 (b), the fluorescence lifetime of the samples is presented. It is better to obtain a longer fluorescence lifetime. This is due to the fact that samples with a longer fluorescence lifetime tended to have a more extended sample lifetime. The fluorescence lifetime of P597 was 5.71 ns, which was the longest fluorescence lifetime within these 6 samples. Second, LumO had a fluorescence lifetime of 4.16 ns. The third place belonged to RhB (3.99 ns). It was then followed by Rh6G (3.44 ns). The fluorescence lifetime of DCM2 and DCJTB were nearly the same, which were 2.41 ns and 2.38 ns, respectively.

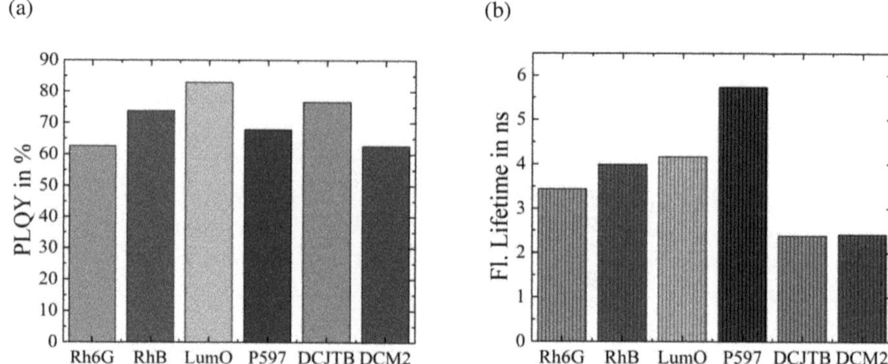

Figure 5.14: (a) PLQY and (b) fluorescence lifetime of the sample doped with different dyes.

The VSL method was applied to compare the optical gain of the samples with different dyes. The pump energy was set at $122.40\,\mu J/cm^2$. The result is illustrated in Figure 5.15 and Figure 5.16. The maximum value of optical gain of each dye was compared here. Among the samples, the P597 doped sample had the highest optical gain. The spectrum optical gain of P597 was also the widest. Meanwhile, the bandwidth of the LumO doped sample was the smallest compared to others. The optical gain spectra of Rh6G and RhB were nearly the same because both had the same basic chemical structure. However, the optical gain spectrum of the RhB doped sample was much red-shifted. As DCJTB and DCM2 had the same basic chemical structure, the optical gain spectrum bandwidth and the measured optical gain were nearly identical.

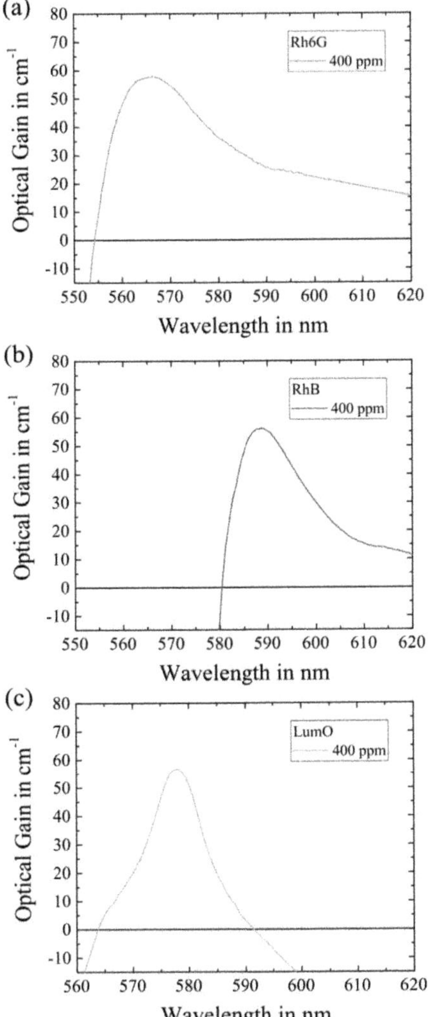

Figure 5.15: Optical gain of the sample doped with different dyes ((a) Rh6G, (b) RhB and (c) LumO) at a pump density of 122.40 µJ/cm^2.

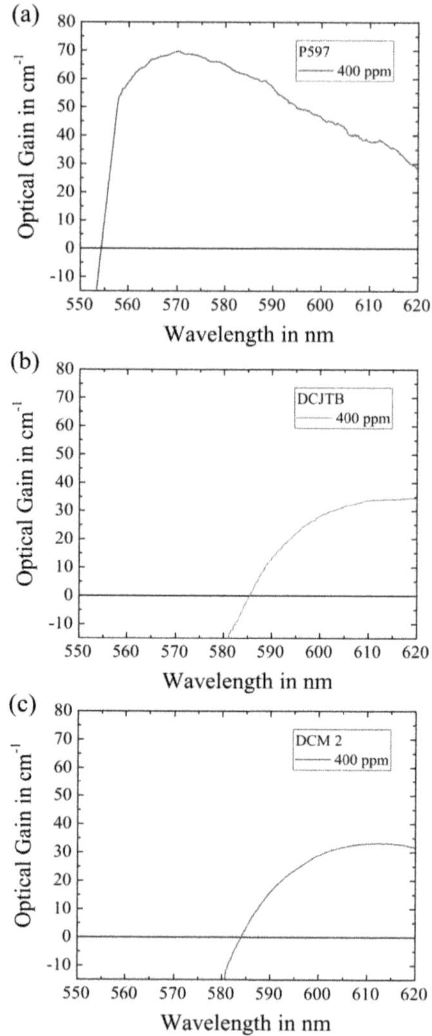

Figure 5.16: Optical gain of the sample doped with (a) P597, (b) DCJTB and (c) DCM2 at a pump density of 122.40 μJ/cm^2.

5.4 Lasing Properties of Different Designs

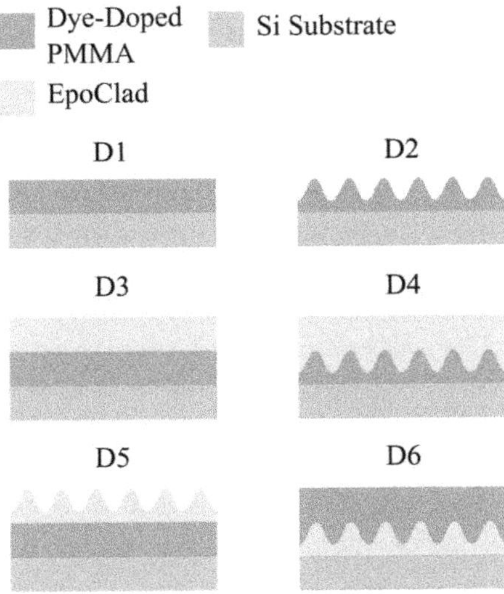

Figure 5.17: Six different designs of sample in this research.

There are different designs that can be applied to fabricate a thin-film laser. To find out which sample model was most practicable with good lasing properties, six different designs as shown in Figure 5.17 were established and tested. For comparison of samples with different designs, Rh6G was selected as the dye-dopant for PMMA because Rh6G is commonly seen in literature to dope in PMMA for fiber laser production [133–135, 143–145]. The concentration was fixed at 400 ppm because the sample doped with 400 ppm of Rh6G had the most significant stimulated emission cross section compared to other concentrations as shown in Chapter 5.2. If the grating structure was in-used, 380 nm grating period was applied according to the result in Table 5.2.

Figure 5.17 provides an overview of all the six sample designs proposed. In the first design (D1), there was only a layer of dye-doped PMMA without any extra optical structure. Next, a grating structure was introduced into the second design (D2). The third design (D3) and the fourth design (D4) were very identical compared to D1 and D2, respectively. In both D3 and D4, an extra transparent layer of EpoClad was proposed on top of the gain medium. The EpoClad layer functioned as a protecting layer or encapsulation layer. Zhai et. al. recommended that the grating structure should not be directly built into the gain medium [120]. Else, the field of the eigenmode could not be distributed completely throughout the layer. To include such a model, the fifth design (D5) was realised, where the grating structure was assembled on top of the layer of EpoClad instead of the doped PMMA layer. The gain medium layer was sandwiched between the substrate and the EpoClad layer with a grating structure. In the sixth design (D6), an EpoClad layer with a grating structure was generated on the Si-substrate at first. Then, the doped PMMA layer was fabricated on top of the EpoClad layer with the grating structure. This kind of sample design is commonly shown in some literature like [29, 92, 97, 103, 107]. The fabrication produce of these sample designs is shown in Chapter 4.2. The layer thickness was different due to the applied fabrication methods. The layer thickness of the design used is illustrated in Table 5.3. In D1, D3, and D5, the doped PMMA layer had a thickness of 1200 nm. Due to the pressure used in the fabrication of D2 and D4, the thickness of the doped PMMA layer was only 702 nm. In D6, the doped PMMA layer was slightly thicker (1280 nm) because it was spin-coated on top of the uneven surface of the grating structure of the EpoClad layer. On the other hand, the EpoClad layer in D3 had a thickness of 782 nm, when no extra pressure or grating structure was applied during the fabrication. In D4, due to the grating structure on PMMA, in which EpoClad was placed on top, the EpoClad layer became thicker (857 nm). When there was a grating structure on the EpoClad layer like in D5 and D6, the EpoClad layer had a thickness of 765 nm.

Table 5.3: Layer thickness of the proposed sample designs

Sample design	Thickness of PMMA layer /nm	Thickness of EpoClad layer /nm
D1	1200	-
D2	702	-
D3	1200	782
D4	702	857
D5	1200	765
D6	1280	765

Lasing Spectrum of Different Design

The first step to investigate the feasibility of the sample designs as a thin-film laser is to measure the lasing spectrum of the samples. Figure 5.18 presents the spectrum of the tested designs under a pump source at the wavelength of 532 nm before the lasing threshold (0.10 µJ) and the lasing spectrum after the lasing threshold (1.69 µJ). Without any optical structure, D1 had a peak wavelength at 566.51 nm and a FWHM of 9.01 nm. This showed that the sample reached only ASE, as Samuel et. al. suggested that the typical FWHM of ASE is around 10 nm. By introducing a grating structure into D2, the peak wavelength was detected at 566.72 nm, while the FWHM was narrowed to 1.16 nm. This indicates that the grating structure functions as a good resonator for lasing indeed. When a layer of EpoClad was added to the grating structure like in D4, the peak wavelength stayed nearly the same (566.67 nm) and a FWHM of 2.18 nm was observed. The FWHM was widened by adding an EpoClad layer to the grating structure. This means that the additional EpoClad layer does not provide further contribution when there is already a grating structure on the gain material. On the other hand, there was only an EpoClad layer on top of gain material like in D3 without grating structure. The peak wavelength was determined at 565.73 nm and the FWHM was 2.90 nm. This shows that the EpoClad layer helps in constricting FWHM, but it is not as efficient as the grating structure. In D5, the grating structure was proposed on the EpoClad layer. The dye-doped PMMA film was sandwiched between the substrate and the EpoClad with a grating structure on top of it. By using D5, the peak wavelength was defined at 575.31 nm and the FWHM was 1.05 nm. This combination of the grating structure and the EpoClad layer demonstrated the best result of FWHM. By switching the EpoClad layer with a grating structure and the gain material layer like in D6, the peak wavelength was evaluated at 577.40 nm and the FWHM was 1.97 nm. In D6, the FWHM became slightly bigger. This was because the gain material was in direct contact with the excitation source without any additional grating structure or EpoClad layer, which meant that the light emission could not be constricted within the gain medium for light amplification.

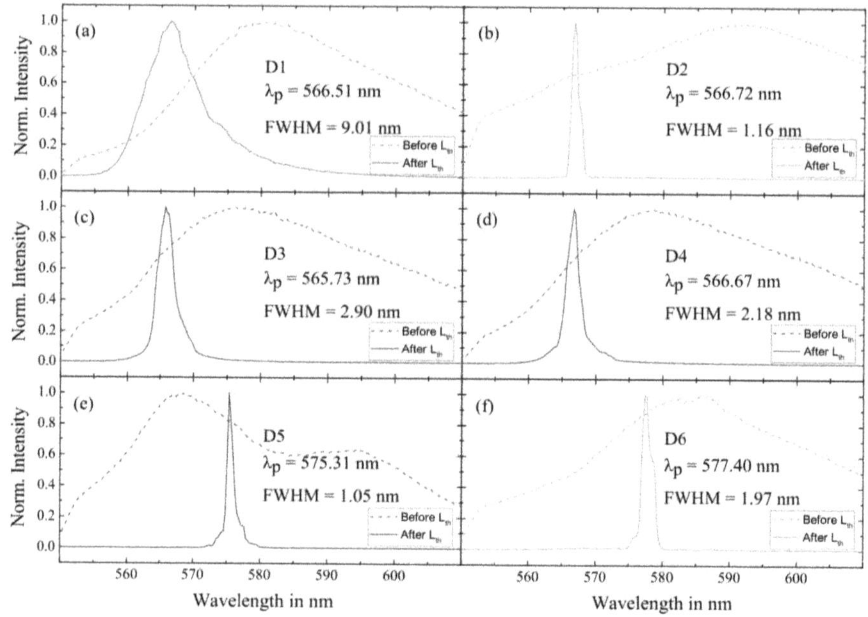

Figure 5.18: Lasing spectrum of different sample designs.

According to [120], Figure 5.19 is depicted, which is a sketch to better understand the distribution of eigenmode in the gain medium. Zhai et al. proposed that the distribution of eigenmode was not properly in the active layer when the grating structure was directly built on the gain material. Based on this theory, the distribution of eigenmode in D1, D3, and D5 was much neatly compared to D2, D4, and D6. This could be a reason, why the FWHM of D2, D4, and D6 was not so confined compared to D5. D5 had not only the comprehensive distribution of eigenmode within the gain medium, but also the EpoClad layer with a grating structure on top of it, which aided in keeping the light mode within the gain medium for light amplification. Therefore, the FWHM of D5 was the smallest among the samples.

A grating structure was introduced in D2, D4, D5, and D6. The peak wavelengths of D2 and D4 were nearly identical, which was about 566 nm. On top of that, D5 had a peak wavelength of 575.31 nm. By using D6, the peak wavelength was slightly shifted to 577.40 nm. There were in total around 10 nm of wavelength shift. The reason was due to the thickness of the layer of doped PMMA. D2 and D4 had the same thickness of the doped PMMA layer (702 nm). Therefore, both samples had the same peak wavelength. Meanwhile, D5 and D6 had a thicker layer thickness of doped PMMA, which were 1200 nm and 1280 nm, respectively. Their peak wavelengths were measured at a longer wavelength. Therefore, it could be assumed that the peak wavelength was longer with increasing layer thickness. When the layer thickness is thicker, there are more dye molecules in the gain medium. This induces a higher possibility of reabsorption within the dye molecules themselves. The re-emission of the reabsorbed photons can shift the peak wavelength to a longer wavelength. This indicates that the peak wavelength can be controlled by the thickness of the gain material, which is also a common technique of lasing tuning in literature [29, 95, 102, 106].

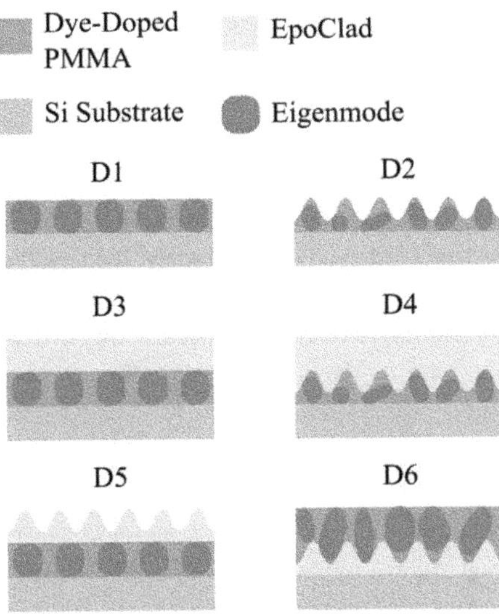

Figure 5.19: A sketch of the distribution of eigenmode along the gain medium according to [120].

Lasing Threshold of Different Design

One of the characteristics of a laser is that the output intensity is amplified upon a specific pump energy. This turning point is named the lasing threshold. A lower lasing threshold is much desired because less energy is needed to induce lasing property. To find out the lasing threshold of the suggested sample designs, the samples were excited with a series of pump energy. The result is illustrated in see Figure 5.20. Among the samples, D1 had the lowest lasing threshold, which was 0.131 µJ, because there was no extra optical structure. By introducing a grating structure to the design like in D2, the lasing threshold was increased to 0.271 µJ. The reason was due to the layer thickness of the gain medium. The thickness of the gain material layer was relatively smaller compared to other samples. This meant that there was less dye in the gain medium. Therefore, more energy was needed to pump the gain material to achieve lasing. In D3, the lasing threshold was slightly higher than D1, which had a value of 0.206 µJ, even though D1 and D3 had the same thickness of the doped PMMA layer. This indicated that EpoClad might absorb some pump energy which caused that much energy was needed for excitation. In D4, there was not only a thinner layer of gain material but also an additional layer of EpoClad. Therefore, the lasing threshold increased to 0.356 µJ. The lasing threshold was affected by the thin dye-doped PMMA layer and the EpoClad layer. In D5, the thickness of gain material was the same as D1 and D3, while an EpoClad layer with a grating structure on top of the doped PMMA was assembled on the doped PMMA layer. The detected lasing threshold was 0.340 µJ. This was caused by the application of the combination of EpoClad and grating structure. In D6, the thickness of doped PMMA was the thickest. However, the lasing threshold was still high, which was 0.353 µJ. The sample with the D6 design was in direct contact with the excitation source. Consequently, the eigenmode could not be confined within the gain medium, which could enhance unwanted optical losses. So, more energy was needed to excite the gain material for the purpose of a laser. Here, it can be deduced that extra optical structures like the grating structure, the EpoClad layer, and its combination can cause an increase of the lasing threshold, even though optical structure helps in narrowing the FWHM.

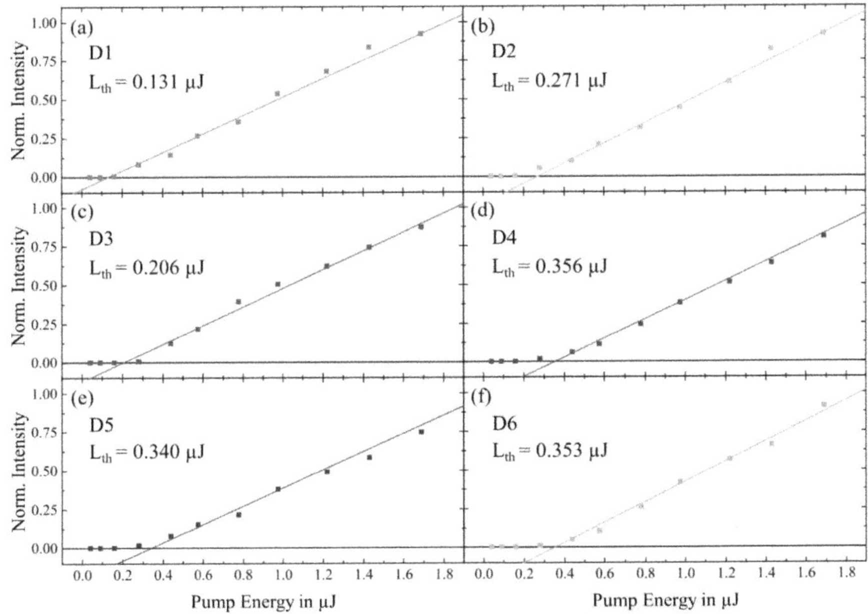

Figure 5.20: Lasing threshold of different sample designs.

PER of Different Design

Polarisation extinction ratio (PER) is one of another laser properties, which compares the ratio between the TE and TM mode. The ratio was calculated by using Equation 4.5. The PER of the samples with the suggested sample structures by using an excitation energy of 1.69 µJ was investigated, while the result is illustrated in Figure 5.21 (a). Among the samples, D1 had the lowest PER, which was only 21.30 dB. By introducing only a grating structure to the sample like in D2, the PER raised to 24.57 dB. The PER stayed nearly the same (24.54 dB), when only an extra EpoClad layer was implemented like in D3. This proves that the grating structure or the EpoClad helps in improving the PER. With the combination of the grating structure and the EpoClad layer, the PER increases further to 28.12 dB in D4 and 30.12 dB in D5, respectively. Therefore, it can be deduced that the combination of grating structure and EpoClad layer can further enhance the PER. The PER of D5 was slightly higher than D4, which was caused by the neat eigenmode distribution of the sample design. Nevertheless, in D6, the PER was only 23.49 dB, even though both grating structure and EpoClad were applied. This result indicates that the sequence of the layer of the gain medium was important. In D6, there was direct contact between the gain medium and the pump source, the PER might therefore be worsened because the light mode could not be properly constricted in the gain medium.

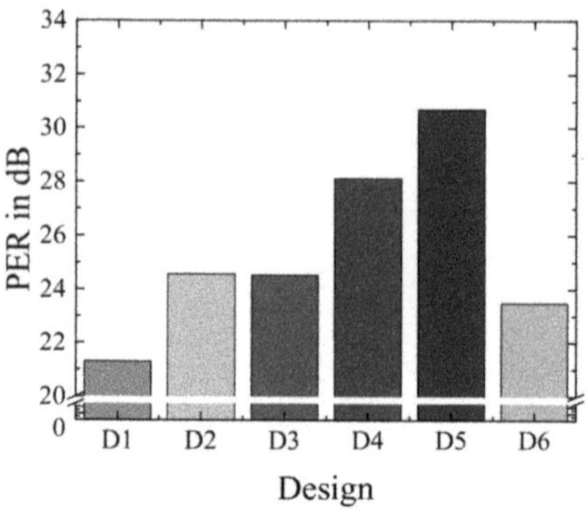

Figure 5.21: PER of different sample designs.

Output Energy of Different Design

Besides, the output energy of the samples is another essential property to be examined, to find out the effect of the sample designs on the laser conversion efficiency. Figure 5.22 shows the output energy of the sample with different sample designs by using a pump energy of 1.69 μJ. Without any aids from additional optical structure, D1 had an output energy of 3.63 nJ. In D2, the output energy was only 1.38 nJ because the thickness of the doped PMMA was smaller compared to D1 due to the fabrication method. When the doped PMMA layer was thinner, there were fewer dye molecules for the contribution of output energy. By adding an EpoClad layer on the grating structure in D4, the output energy was increased to 6.00 nJ, even though the thickness of the doped PMMA was the same as D2. This demonstrates that the EpoClad layer on gain material can help in strengthening the output energy by keeping the emission of gain material in the gain medium for better light amplification. In D3, an output energy of 8.56 nJ was determined. It was a bit higher than D1 and D2. This was because the gain material layer was the same as D1, in addition to the help of the EpoClad layer. In D5, the output energy was highest, which was 12.21 nJ. In D5, the doped PMMA layer thickness was the same as D1 and D3. Besides, the neat eigenmode distribution in the gain medium plus the combination of the grating structure and the EpoClad layer enhanced the output energy of D5. In D6, due to the probability of disorganised eigenmode distribution and the direct interaction with excitation energy, the output energy became lower (8.82 nJ).

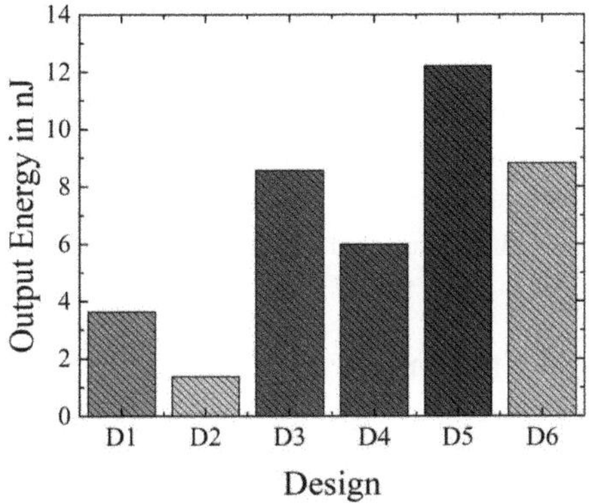

Figure 5.22: Output energy of different sample designs.

Based on the results, D5 was selected for further research. Even though the lasing threshold of D5 was not the smallest among the sample models, it had the highest output energy and PER. On top of this, the FWHM of D5 was the smallest among the samples. The layer thickness of Rh6G doped PMMA and EpoClad was further modified to improve the performance of the device, which was discussed later on.

5.4.1 Device Improvement

As mentioned before, D5 was the best sample design among the proposed structures. The layer thickness of dye-doped PMMA and EpoClad was adjusted to assure the device performance of D5. With the changes in the layer thickness of doped PMMA and EpoClad, the output energy was mainly focused. In addition, other lasing characteristics like lasing spectrum and lasing threshold were also observed here.

Thickness Variation of EpoClad

In general, there were two methods to change the thickness of the EpoClad layer: First, the viscosity of EpoClad could be controlled through dilution with Gbl. Second, the layer thickness of EpoClad could be altered by using a different speed of spin-coating.

EpoClad Layer Thickness by Dilution

Figure 5.23 presents the effect of the ratio of EpoClad to Gbl on the layer thickness, by adding more Gbl to dilute the EpoClad. For this experiment, the speed of spin-coating was kept at 4000 rpm. It is noticed that the layer thickness of EpoClad reduced exponentially when the portion of Gbl increased. When the EpoClad was 75wt%, the layer thickness expressed 4306 nm. By using 67wt% of EpoClad, the layer thickness was further decreased to 2408 nm. Next, the EpoClad had a layer thickness of 765 nm, when Epoclad was 50wt%. When the portion of Gbl was double more than EpoClad (33wt%), the layer thickness of EpoClad was only 335 nm. The portion of Gbl could be further increased for dilution, while the layer thickness of EpoClad tended to reach a minimum. With 25wt% and 20wt% EpoClad, the layer thicknesses were 199 nm and 146 nm, respectively.

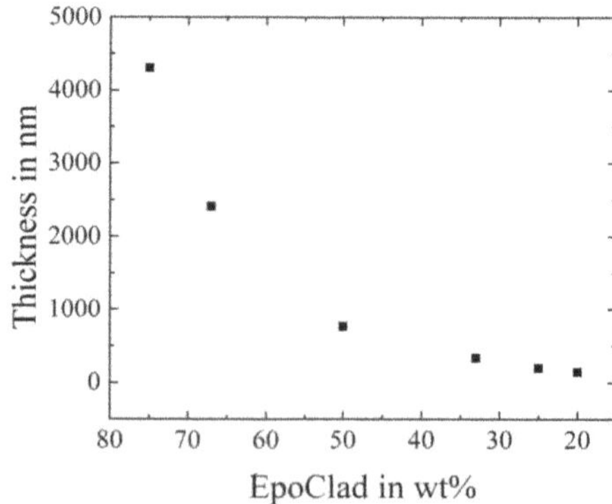

Figure 5.23: Thickness of EpoClad layer through dilution by using gamma-butyrolactone (Gbl).

Output Energy of EpoClad Thickness Variation by Dilution

The measurement of output energy of the sample was observed at a pump energy of 1.69 µJ (see Figure 5.24). By reducing the ratio of EpoClad to 75wt% to 50wt%, the output energy increased from 7.66 nJ to 12.21 nJ. An enhancement of output energy was observed which was caused by the decreasing of the EpoClad layer thickness. The reason for this tendency could be that EpoClad might absorb some excitation energy. Therefore, less excitation energy was absorbed when the thickness was smaller which enhanced the output energy. However, when the viscosity of EpoClad was furthered diluted, there was a vast decrease in output energy, which was lower than 3.5 nJ. When the dilution was too high, the EpoClad lost its stickiness for the attachment of the stamp during sample production. As a consequence, the grating structure could not be appropriately formed. So, the output energy was low. Among the samples, 50wt% of EpoClad was the best recipe. Still, the thickness can be controlled by varying the speed of spin-coating, while the result was discussed later.

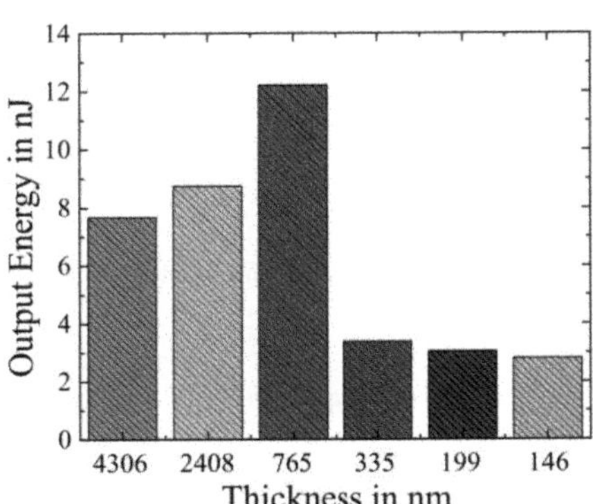

Figure 5.24: Output energy of the sample with various thickness of the EpoClad layer by using the method of dilution.

EpoClad Layer Thickness by Spin-coating Speed

As stated before, the performance of the samples can be changed by alternating the layer thickness EpoClad. 50wt% EpoClad was the best because it had the highest output energy as shown in Figure 5.24. Another method to modify the layer thickness of EpoClad was by adjusting the speed of spin-coating. Figure 5.25 shows the relationship between the speed of the spin-coating and the layer thickness of the EpoClad. With the increasing speed of spin-coating, the layer thickness of EpoClad decreased linearly. When the speed was set at 4000 rpm, the layer thickness of EpoClad expressed 760 nm. The layer thickness of EpoClad was reduced to 712 nm when the spin-coating speed was increased to 4500 rpm. With a spin-coating speed of 5000 rpm, the layer thickness of EpoClad showed 661 nm. The layer thickness of EpoClad was further reduced to 628 nm with a spin-coating speed of 5500 rpm. By increasing the spin-coating speed to 6000 rpm, the layer thickness of EpoClad was minimised to 601 nm. The layer thickness of EpoClad was the smallest among the samples, which had a value of 559 nm, when the speed of spin-coating was appointed at 6500 rpm.

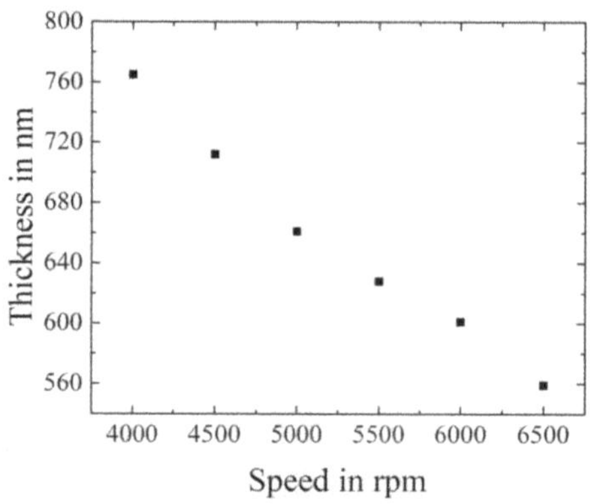

Figure 5.25: EpoClad layer thickness by using different speeds of spin-coating.

Output Energy of EpoClad Thickness Variation by Spin-coating Speed

The output energy of the samples was also investigated at a pump energy of 1.69 µJ. The result is presented in Figure 5.26. With the layer thickness of EpoClad of 765 nm, an output energy of 12.21 nJ is measured. When the layer thickness of Epoclad was further decreased to 712 nm, a higher output energy with a value of 12.63 nm was examined. The output energy of the samples with the EpoClad layer thickness of 661 nm was the highest among the samples, which was 16.43 nJ. The output energy increased when the EpoClad thickness decreased, because the EpoClad layer thickness absorbed some of the excitation energy. With decreasing EpoClad layer, less excitation energy was absorbed, so the output energy became higher. On the other hand, when a much thinner layer thickness of EpoClad was applied, the output energy became lesser. With an EpoClad layer thickness of 628 nm, an output energy of 13.68 nm was obtained. The output energy was minimised to 12.41 nJ by using the layer thickness of EpoClad of 601 nm. The output energy of the samples with EpoClad layer thickness of 559 nm was the smallest compared to others, which was 7.70 nJ. The output energy decreased upon a certain speed of spin-coating. When the spin-coating speed was too high, there was uncertainty on the layer thickness, which could cause a defect in the grating structure. Therefore, the output energy was worsened. Accordingly, the best prescription for the production of the EpoClad layer was that the 50wt% EpoClad was spin-coated at 5000 rpm, which had a layer thickness of 661 nm.

In short, by using different speeds of spin coating, the layer thickness of EpoClad could be controlled, which brought more promising results. The layer thickness of EpoClad had a range from about 600 nm to 800 nm. At the same time, most of the samples had an output energy of more than 10 nJ. To understand more about the effect of the layer thickness of EpoClad on the performance of the samples, other lasing characteristics like lasing spectrum and lasing threshold were checked.

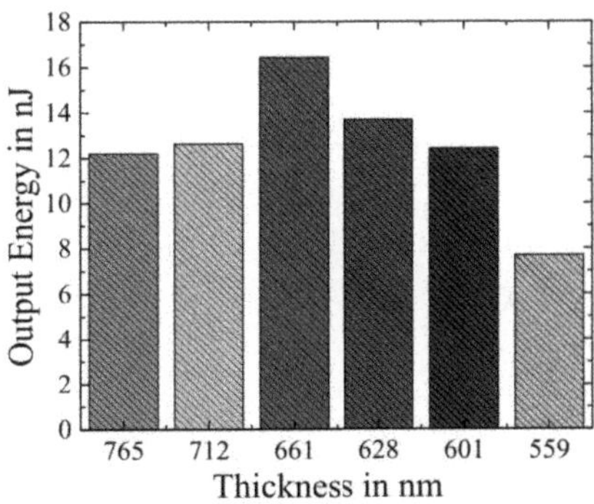

Figure 5.26: Output energy of the sample with different EpoClad layer thicknesses under the effect of various spin-coating speeds.

Lasing Spectrum of EpoClad Thickness Variation

The lasing spectrum of the samples was measured, in which the samples had different EpoClad layer thicknesses (see Figure 5.27). By using the EpoClad layer thickness of 765 nm, a spectrum with a peak wavelength at 575.31 nm and a FWHM of 1.05 nm were measured. The peak wavelength of the spectrum was slightly shifted to 575.15 nm with the implementation of the layer thickness of EpoClad of 712 nm, while the FWHM of the spectrum was 1.00 nm. When the applied EpoClad layer thickness was further decreased to 661 nm, the peak wavelength was found at 575.05 nm and the FWHM was further minimised to 0.83 nm. With the application of 628 nm thick of EpoClad, the peak wavelength of the spectrum was shifted to 574.01 nm, while the FWHM was slightly increased to 1.03 nm. A peak wavelength of 573.23 nm and a FWHM of 1.11 nm were seen by using the EpoClad with a layer thickness of 601 nm. When the 559 nm thick of EpoClad was employed, a peak wavelength was found at 572.97 nm and a FWHM of 1.20 nm was recognised. As a short outline, by using EpoClad with the layer thickness at the range from 559 nm to 765 nm, there was only a 2 nm shift of the peak wavelength from about 573 nm to 575 nm. The result agrees with Bonal et.al., who suggested that the peak wavelength moved to a slightly shorter wavelength [32] with decreasing residue layer of the grating structure. At the same time, the FWHM had a value of around 1.00 nm ± 0.20 nm, which was not strongly affected by the thickness of the EpoClad layer.

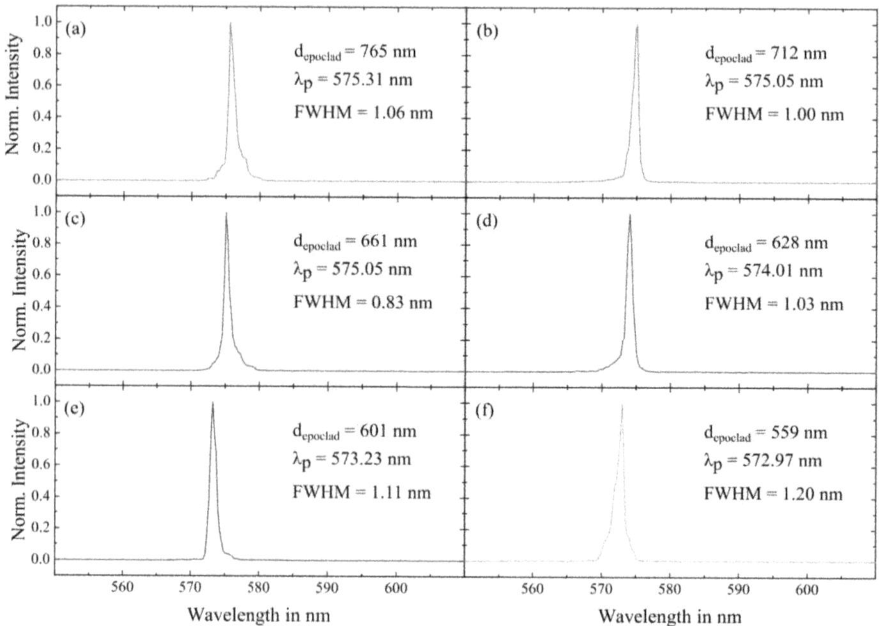

Figure 5.27: Lasing spectrum of the samples with different layer thicknesses of EpoClad.

Lasing Threshold of EpoClad Thickness Variation

In addition to lasing spectrum, the changes in the layer thickness of EpoClad would affect the lasing threshold, which provided the information about the needed pump energy to induce laser. The result is shown in Figure 5.28. When a layer thickness of EpoClad of 765 nm was used, the lasing threshold was found at 0.341 µJ. The lasing threshold decreased to 0.324 µJ with the usage of 712 nm thick of EpoClad layer. When the layer thickness of EpoClad was further reduced to 661 nm, a lasing threshold of 0.311 nJ was measured. When the EpoClad had a layer thickness of

628 nm, the lasing threshold implied 0.290 µJ. With the application of 601 nm thick EpoClad, the lasing threshold became 0.244 µJ. The lasing threshold with a value of 0.235 µJ was the smallest among the samples when the applied EpoClad layer thickness was 559 nm, which was also the thinnest compared to others. In short, when the layer thickness of EpoClad became smaller, the lasing threshold was reduced on a small scale. This is because less energy was absorbed by the EpoClad layer with decreasing EpoClad thickness, so less energy was required for lasing activation.

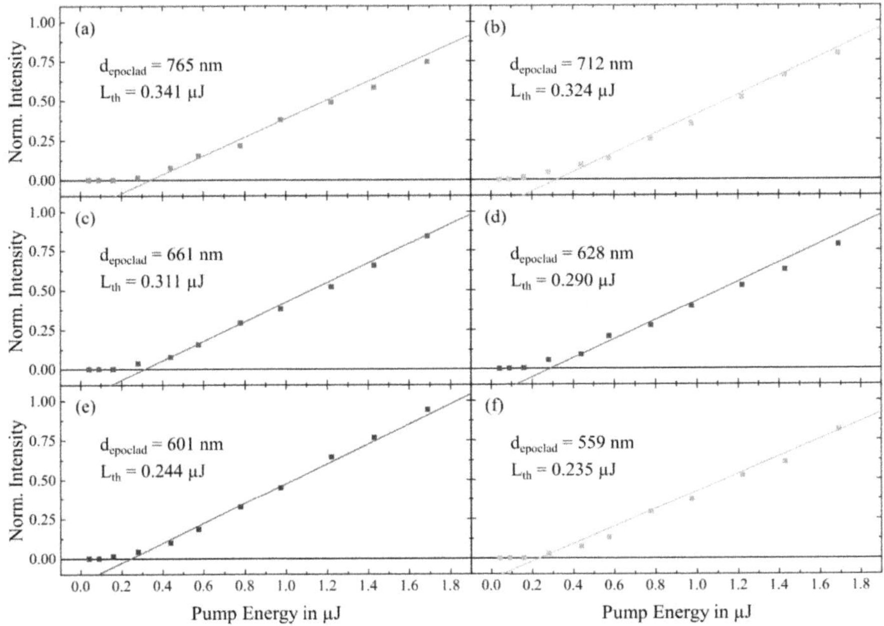

Figure 5.28: Lasing threshold of the sample with different layer thicknesses of EpoClad.

For further research, the recipe of the EpoClad layer was kept at 50wt% EpoClad by using 5000 rpm of spin-coating speed. It had a layer thickness of 661 nm. The output energy was the highest (16.43 nJ) among the samples. Besides, the FWHM of the spectrum (0.83 nm) was the smallest. In spite of the fact that the lasing threshold (0.311 µJ) was not the smallest, it was still comparable to the others.

Thickness Variation of PMMA

The doped PMMA was the gain medium of the laser of the sample design. By modifying the layer thickness of the gain medium, the resulting lasing properties would be affected. The concentration of PMMA in butanone here was kept at 7.5 wt% with respect to butanone. By using a lower concentration of PMMA to butanone, the optimal layer thickness based on simulation (1200 nm) could not be achieved. On the other hand, the sample surface became inhomogeneous, when the PMMA concentration was higher. By keeping 7.5 wt% PMMA concentration, it can be assured that the distribution of Rh6G dye molecule in the PMMA matrix was constant. To investigate the effect of the layer thickness of the gain medium, the layer thickness was regulated by using different speeds of spin-coating.

PMMA Layer Thickness by Spin-coating Speed

Figure 5.29 presents the influence of the speed of spin-coating on the layer thickness of PMMA. The thickest layer of PMMA of 1446 nm was obtained, when the speed of spin-coating was set at 250 rpm. No flat thin film could be achieved if the lower speed was applied. When the speed of spin-coating was increased to 350 rpm, the thickness of the PMMA layer became 1200 nm. A 1002 nm thick PMMA layer was measured with the usage of 500 rpm of spin-coating speed. The PMMA layer was further decreased to 846 nm, when the speed of spin-coating was further increased to 800 rpm. With a speed of spin-coating of 1000 rpm, the thickness of the PMMA layer expressed 753 nm. The PMMA layer was the thinnest among the samples, which showed a value of 606 nm, when the speed of spin-coating was 2000 nm. As a short summary, the result shows a trend that the layer thickness of PMMA decreased linearly with increasing speed of spin-coating.

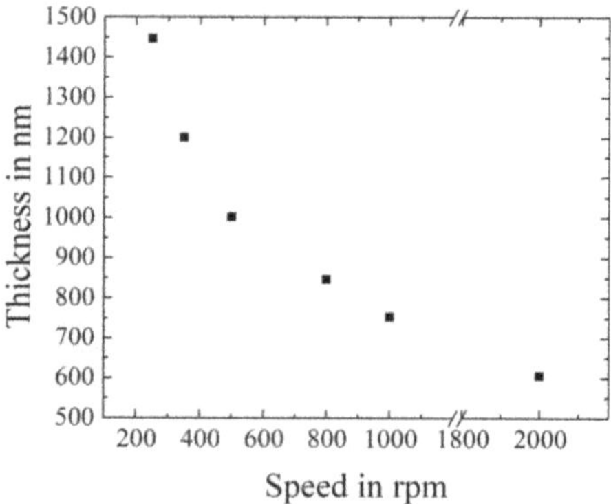

Figure 5.29: Thickness of PMMA layer by using different speeds of spin-coating.

Output Energy of PMMA Thickness Variation

With the changes in the thickness of the Rh6G doped PMMA layer, the influence on output energy was measured at a pump energy of 1.69 µJ (see Figure 5.30). When the thickest PMMA layer (1446 nm) was applied, an output energy of 11.61 nJ was measured. The highest output energy was evaluated, which was 16.43 nJ, when the PMMA layer was 1200 nm. When the PMMA layer was decreased to 1002 nm, the output energy was reduced to 10.01 nm. With the application of 846 nm of the PMMA layer, the output energy was only 4.00 nJ. The output energy became much less, which was 1.99 nJ, when the applied PMMA was 753 nm. With the usage of 606 nm thick of PMMA layer, the output energy was the smallest among the samples, which was only 0.59 nJ. It can be deduced that the output energy reduces, when the layer thickness is smaller. The reason could be that there were fewer dopants in the thinner gain material layer, which could contribute to the output energy. However, when the doped PMMA layer was comparably too thick (>1200 nm), the output energy decreased. When a thick layer was produced by spin-coating, an inhomogeneous surface tends to occur, which could worsen the output energy.

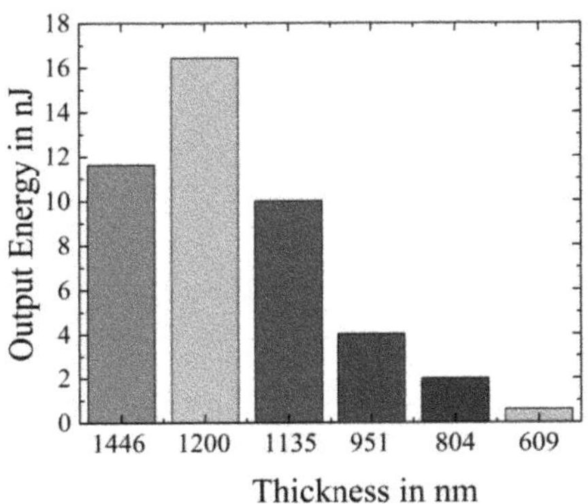

Figure 5.30: Output energy of the sample with different PMMA thicknesses.

Lasing Spectrum of PMMA Thickness Variation

The spectrum of the samples with different thicknesses of the PMMA layer was investigated at a pump energy of 1.69 µJ, which is shown in Figure 5.31. When the PMMA layer with a thickness of 1446 nm was applied, two peak wavelengths are detected, which were 575.40 nm and 575.83 nm. Meanwhile, the FWHM had a value of 2.74 nm. When the used PMMA layer was 1200 nm, the peak wavelength was found at 575.05 nm and it had a FWHM of 0.83 nm. With the usage of

1002 nm thick of PMMA layer, the spectrum had a peak wavelength of 574.01 nm and a FWHM of 0.86 nm. The FWHM became slightly bigger with a value of 0.95 nm, when the PMMA had a layer thickness of 846 nm, in which the peak wavelength was recognised at 572.45 nm. With the application of a PMMA layer thickness of 753 nm, a peak wavelength of 570.89 nm was spotted, while the FWHM was 0.98 nm. There was a significant shift of peak wavelength when the applied PMMA layer thickness was 606 nm, in which the peak wavelength was 563.33 nm. However, the FWHM widened, which consisted of a value of 32.18 nm.

When the PMMA layer was too thick (1446 nm), there were two peak wavelength seen because a thick layer allowed other light mode to propagate. On the other hand, the FWHM became too wide when the PMMA layer was too thin (563.33 nm). It meant that the sample did not even reach the state of ASE. Single-mode laser was only detected when the PMMA layer thickness ranged from 753 nm to 1200 nm, because only one peak wavelength was measured. Besides, there was about 4 nm red-shift with increasing PMMA layer thickness. When the layer thickness was higher, there were more dye molecules in the gain medium. This meant that the possibility of reabsorption of dye molecules was higher. The re-emission of the reabsorbed photons caused the red-shift

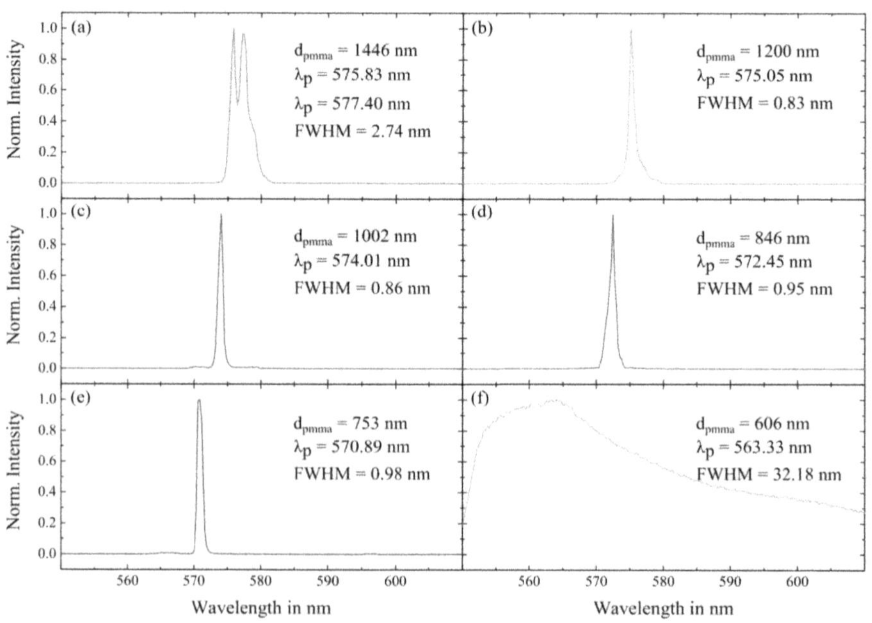

Figure 5.31: Lasing spectrum of the sample with different PMMA layer thicknesses.

Lasing Threshold of PMMA Thickness Variation

One of the aims of this work was to achieve a sample with a single-mode laser. This meant that the lasing spectrum should have only a single peak wavelength with narrowing FWHM. As seen from Figure 5.31, the sample with a 1446 nm thick PMMA layer did not count as a single-mode laser because two peak wavelengths are observed. Additionally, the sample with 606 nm thick PMMA layer was also not considered as a laser due to the fact that the FWHM was too wide. To understand the effect of the thickness of the PMMA layer on lasing threshold, only the rest four samples were further tested, while the result was presented in Figure 5.32.

By having a PMMA layer thickness of 1200 nm, the lasing threshold implied a value of 0.311 µJ. When the thickness of the PMMA layer was decreased to 1002 nm, a slightly higher lasing threshold with a value of 0.335 µJ was detected. With a PMMA layer thickness of 846 nm, the lasing threshold was 0.350 µJ. The lasing threshold of the sample with the PMMA layer thickness of 753 nm was the highest compared to others, which had a value of 0.389 µJ. As a short summary, with decreasing PMMA layer thickness, the lasing threshold increased to a small extent. This is due to the fact that there is less amount of dopants when the layer thickness becomes smaller. Therefore, more energy is needed for laser activation.

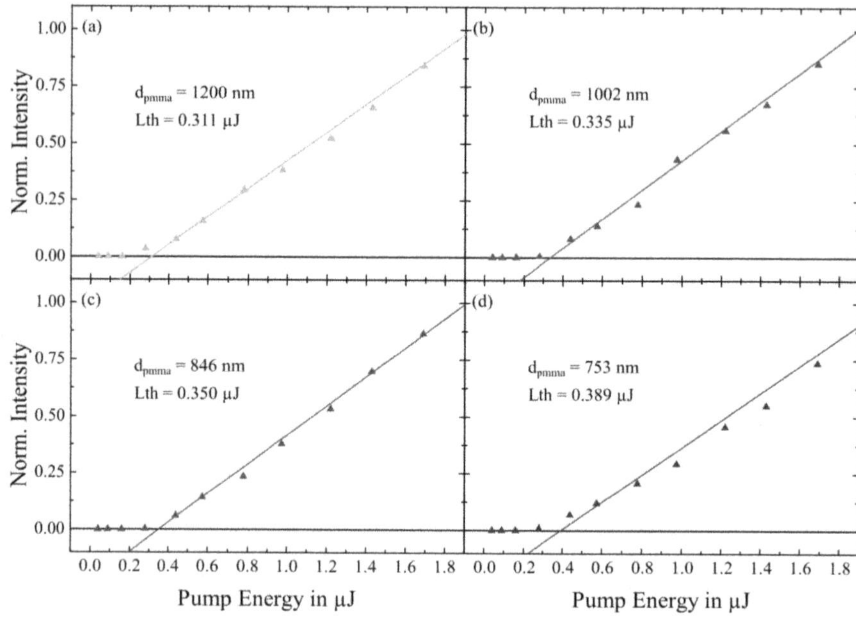

Figure 5.32: Lasing threshold of the sample with different layer thicknesses of PMMA.

After alternating the thickness of the EpoClad layer and the PMMA layer of D5, it was found out that the sample with EpoClad layer thickness of 661 nm and PMMA layer thickness of 1200 nm was the best among the samples. It had a peak wavelength at 575.05 nm. Furthermore, the FWHM of the spectrum was 0.83 nm. Besides, the measured lasing threshold was 0.311 μJ. For further research, D5 with following properties was applied: the thickness of the EpoClad layer was kept at 661 nm by using 50wt% of EpoClad and a 5000 rpm spin-coating speed. Meanwhile, the PMMA layer thickness was set at 1200 nm by using 7.5 wt% PMMA concentration with respect to butanone and a spin-coating speed of 350 rpm.

5.5 Lasing Tuning

One of the criteria of a laser is that the lasing properties are strongly affected by changes of the resonator and the gain medium. With an alternation of the resonator or the gain medium, the laser can be tuned. Laser tuning is an interesting topic, in which the peak wavelength of the samples can be manipulated. There are several methods to reach this purpose. In this work, three different methods were applied. First, the concentration of the dopant in the samples was varied. Second, the applied Bragg grating period was manipulated. Third, different dyes were doped into the samples to check the feasibility of other laser dyes in PMMA. As discussed before, the D5 sample model with 661 nm thick of EpoClad layer and 1200 nm thick of PMMA layer performed the best result among the designs. Therefore, samples with these properties were fabricated for the purpose of lasing tuning.

5.5.1 Dye Concentration Variation

In Chapter 5.4, it was proven that Rh6G doped PMMA thin-film with grating structure demonstrated lasing properties like a small FWHM and a clear lasing threshold. To fulfill the lasing criteria, the effect of the changes of the gain medium and the resonator on lasing properties should be checked. The first method used here was to manipulate the dye concentration. As stated in Chapter 5.2, only three concentrations of Rh6G could be doped into PMMA, which were 100 ppm, 200 ppm, and 400 ppm. The grating period was kept at 380 nm. By changing the Rh6G concentration, the laser was tuned. The lasing properties such as lasing spectrum, lasing threshold, PER, and output energy were measured.

Lasing Spectrum of Dye Concentration Variation

Figure 5.33 shows the influence of different Rh6G concentrations on the spectrum before the lasing threshold (0.10 μJ) and the lasing spectrum after the lasing threshold (1.69 μJ). When the 100 ppm Rh6G concentration was employed, the lasing spectrum had a peak wavelength of 563.07 nm with a FWHM of 10.85 nm. The FWHM was too big, which symbolised that the sample only reached ASE without lasing according to [142]. When the Rh6G concentration was increased to 200 ppm, the peak wavelength of the lasing spectrum was shifted to 569.06 nm, while the FWHM was 1.05 nm. With the implementation of 400 ppm Rh6G concentration, the lasing spectrum consisted of a peak wavelength at 575.05 nm. The FWHM of the spectrum of 100 nm was relatively wide compared to others. At the same time, about 6 nm of red-shift was observed between the samples with 200 ppm concentration and 400 ppm concentration. By using higher Rh6G concentration, there were more Rh6G dye molecules in the layer thickness of the gain medium, which led to a higher possibility of reabsorption. The re-emitted of reabsorbed photons could induce the red-shift. According to Equation 3.11, the FWHM can be affected by the optical gain. The 400 ppm Rh6G doped sample had a higher spectral optical gain ($44.94\,cm^{-1}$) than the 200 ppm doped sample ($38.20\,cm^{-1}$). This shows that the FWHM becomes smaller with a higher optical gain.

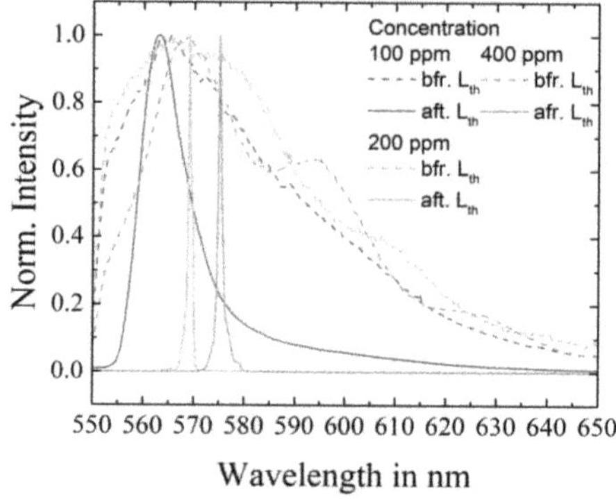

Figure 5.33: Lasing spectrum of the sample with different Rh6G concentrations: 100 ppm, 200 ppm, and 400 ppm.

Lasing Threshold of Dye Concentration Variation

Another lasing property is the lasing threshold that can also be affected by the dye concentration in the gain medium. By changing the Rh6G concentration, the resulted lasing threshold is displayed in Figure 5.34. When the sample had a 100 ppm Rh6G concentration, there was no lasing threshold detected because there was no turning point where the output intensity was strengthened. This confirmed that the lasing state was not reached, which is due to the reason that the concentration was too low. The number of dye molecules in 100 ppm was too small for lasing activation. Meanwhile, the sample with 200 ppm Rh6G concentration had a lasing threshold of 0.530 µJ. The lasing threshold decreased to 0.311 µJ, when the sample had a Rh6G concentration of 400 ppm. According to Equation 3.7, the lasing threshold was affected by the stimulated emission cross section. The higher the stimulated emission cross section, the lower the lasing threshold. The spectral stimulated emission cross section can be calculated based on Equation 3.4. The 400 ppm Rh6G doped sample had a higher spectral stimulated emission cross section ($2.87 \times 10^{-17}\,cm^{-2}$) than the 200 ppm Rh6G doped sample ($2.34 \times 10^{-17}\,cm^{-2}$). Therefore, the lasing threshold of the 400 ppm Rh6G doped sample was lower than that of the 200 ppm Rh6G doped sample.

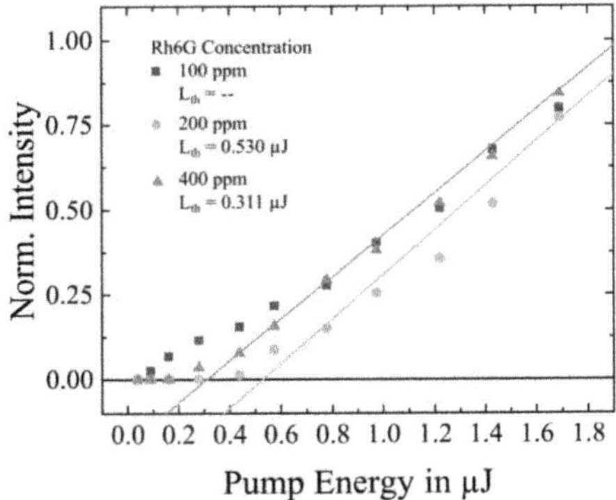

Figure 5.34: Lasing threshold of the sample with three Rh6G concentrations: 100 ppm, 200 ppm, and 400 ppm.

PER of Dye Concentration Variation

In addition to lasing spectrum and lasing threshold, the PER of the samples with varied Rh6G concentrations at a pump energy of 1.69 µJ was also investigated (see Figure 5.35). When the sample had a Rh6G concentration of 100 ppm, a PER of 15.58 dB was determined. The PER increased to 28.94 dB, when the Rh6G concentration was 200 ppm. When the Rh6G concentration was further raised to 400 ppm, the PER was only slightly strengthened to 30.88 dB. It could be assumed that the PER was low when the lasing state was not reached. However, once the lasing state was achieved, the measured PER would be >28 dB, which was not strongly affected by the dye concentration.

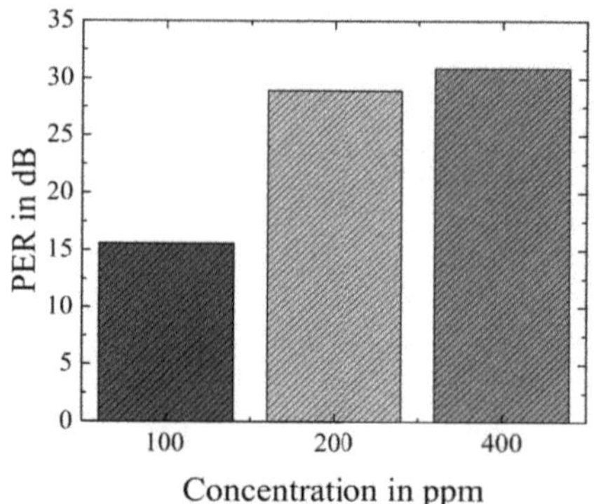

Figure 5.35: PER of the sample with different Rh6G concentrations: 100 ppm, 200 ppm, and 400 ppm.

Output energy of Dye Concentration Variation

Besides, the output energy is a very important characteristic to be studied in order to find out the lasing conversion efficiency, which can be strongly affected by the dopant concentration in the gain medium. Figure 5.36 shows the output energy induced by different Rh6G concentrations at a pump energy of 1.69 µJ. It was very obvious that the sample with 100 ppm had the lowest output energy, which was only 0.22 nJ. When the Rh6G concentration increased to 200 ppm, the measured output energy raised to 6.12 nJ. The output energy of the sample with 400 ppm Rh6G concentration was the highest among the sample, which was 16.43 nJ. Therefore, it can be deduced that the output energy was enhanced with increasing Rh6G concentration. The trend of output energy could be explained by the spectral stimulated emission cross section, which defined the emission efficiency. As the 400 ppm Rh6G doped sample had the highest spectral stimulated emission cross section, the output energy was also the highest.

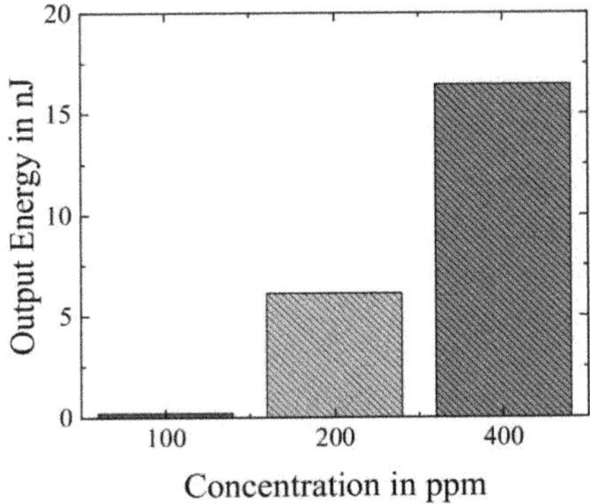

Figure 5.36: Output energy of the sample with different Rh6G concentrations: 100 ppm, 200 ppm, and 400 ppm.

5.5.2 Grating Period Variation

The grating period used in the design structure of D5 functioned as a resonator, which only reflected a specific wavelength for light amplification to generate a laser. In addition to a small FWHM and a clear lasing threshold, the resonator needed to be manipulated to confirm the lasing phenomena. By changing the resonator, the laser could be tuned. Three different grating periods were employed, which are 370 nm, 380 nm, and 390 nm. In this experiment, the Rh6G dye concentration was fixed at 400 ppm.

Lasing spectrum of Grating Period Variation

Figure 5.37 displays the spectrum under the implementation of three different grating periods before the lasing threshold (0.10 µJ) and the lasing spectrum after the lasing threshold (1.69 µJ). By applying the grating period of 370 nm, the lasing spectrum consisted of a peak wavelength of 563.85 nm and a FWHM of 0.70 nm. When the engaged grating period was 380 nm, the peak wavelength of the lasing spectrum moved to 575.05 nm and the FWHM was 0.83 nm. With the usage of the grating period of 390 nm, the peak wavelength of the lasing spectrum shifted to 588.86 nm, while the FWHM was 0.95 nm. As an outline, there was in total an approximate 25 nm shift of wavelength by using these three grating periods. According to Equation 3.13, with a bigger grating period, the peak wavelength would shift to a longer wavelength. Furthermore, the FWHM increased in a small scale when the grating period became bigger. The size of FWHM could be explained by using the optical gain based on Equation 3.11. In Figure 5.38, the spectral optical gain was bigger with a shorter grating period. Therefore, it can be assumed that the FWHM became smaller with a bigger optical gain.

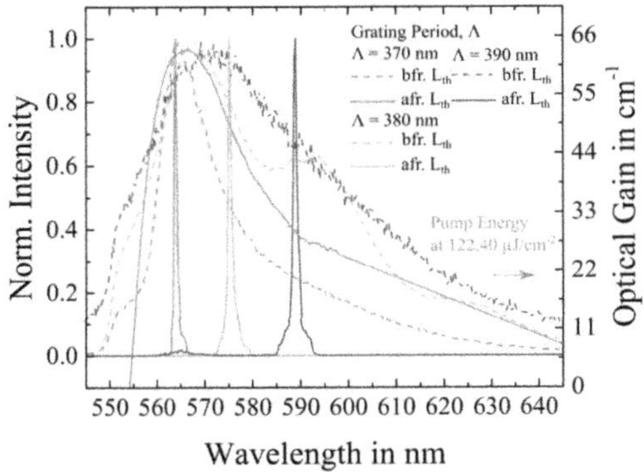

Figure 5.37: Lasing spectrum of the sample with different grating periods (370 nm, 380 nm, and 390 nm) with the spectrum of optical gain in background.

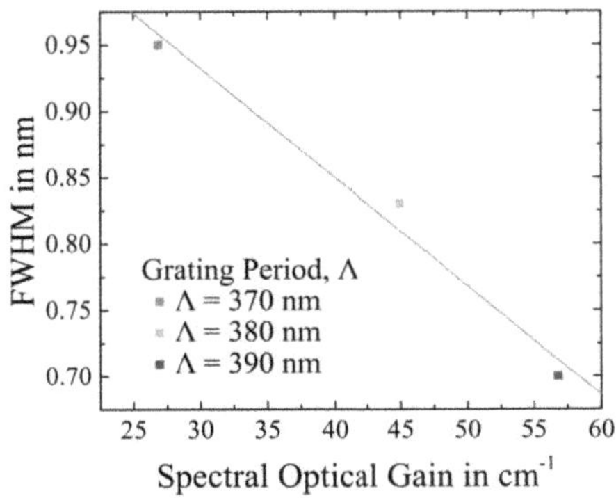

Figure 5.38: Relationship between the spectral optical gain and the FWHM under the consequence of different grating periods.

Lasing Threshold of Grating Period Variation

Since the lasing spectrum is changed, the lasing threshold can be altered as well. Therefore, it is essential to measure the resulting lasing threshold. The result is illustrated in Figure 5.39. By using the grating period of 370 nm, a lasing threshold of 0.156 μJ was recognised. The lasing threshold increased to 0.311 μJ, when the grating period of 380 nm was applied. With the implementation of the 390 nm grating period, the lasing threshold became much higher, which was 0.566 μJ. Thus, it can be deduced that, by using a bigger grating period, the lasing threshold became higher. The lasing threshold could be influenced by the spectral stimulated emission cross section, which could be calculated by using the spectral optical gain based on Equation 3.4. The relationship was shown in Figure 5.40 (b), which showed that the lasing threshold was lower with increasing spectral stimulated emission cross section, while the sample with a smaller grating period had a higher spectral stimulated emission cross section.

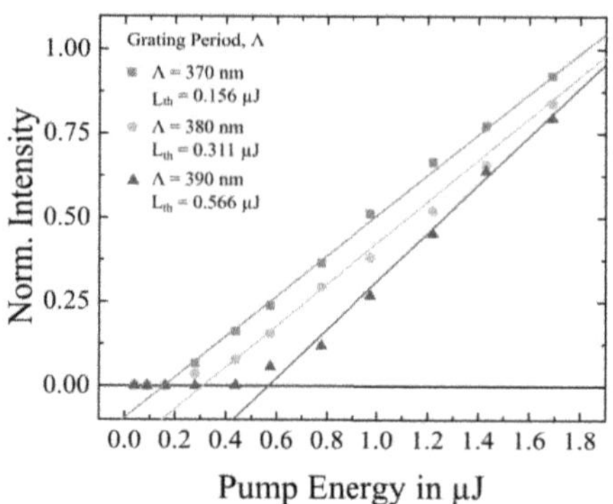

Figure 5.39: Lasing threshold of the sample with different grating periods (370 nm, 380 nm, and 390 nm).

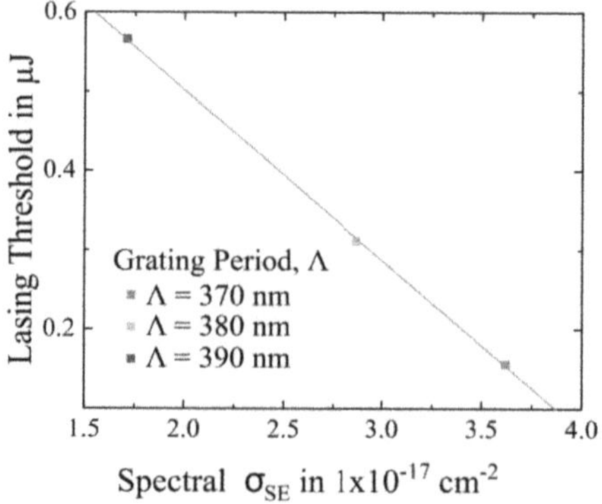

Figure 5.40: Relationship between the spectral stimulated emission cross section and the lasing threshold under the effect of different grating periods.

PER of Grating Period Variation

On top of the lasing spectrum and the lasing threshold, the PER of samples with different grating periods was also examined at a pump energy of 1.69 μJ. The result is demonstrated in Figure 5.41. The PER of the samples with the grating period of 370 nm was nearly identical to the sample with the grating period of 380 nm, which were 30.97 dB and 30.88 dB, respectively. Meanwhile, the PER of the samples with the grating period of 390 nm was slightly smaller compared to others, which was 28.27 dB. It was still comparable to other grating periods. Therefore, it can be understood that the PER was at least 28 dB, despite the employment of various grating periods, as the lasing state was reached.

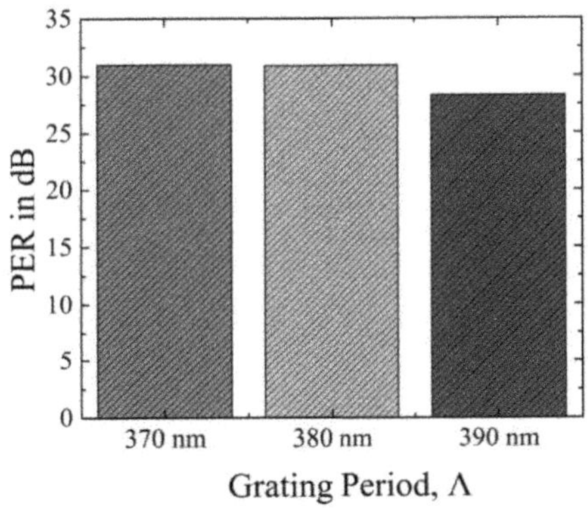

Figure 5.41: PER of the sample with different grating periods (370 nm, 380 nm, and 390 nm).

Output Energy of Grating Period Variation

Just as importantly, the output energy was also determined at a pump energy of 1.69 μJ (see Figure 5.42). When the grating period of 370 nm was applied, an output energy of 17.13 nJ was determined. The output energy decreased slightly to 16.43 nJ with the employment of the grating period of 380 nm. When the sample had a grating period of 390 nm, the measured output energy was only 11.78 nJ. In general, with the implementation of all the tested grating periods, the output energy was over 10 nJ. Stimulated emission cross section is defined as the emission efficiency, which can be used to explain the trend of output energy. The smaller grating periods had a higher spectral optical gain which led to a higher spectral stimulated emission cross section. A higher spectral stimulated emission cross section could induce a higher output energy (see Figure 5.43).

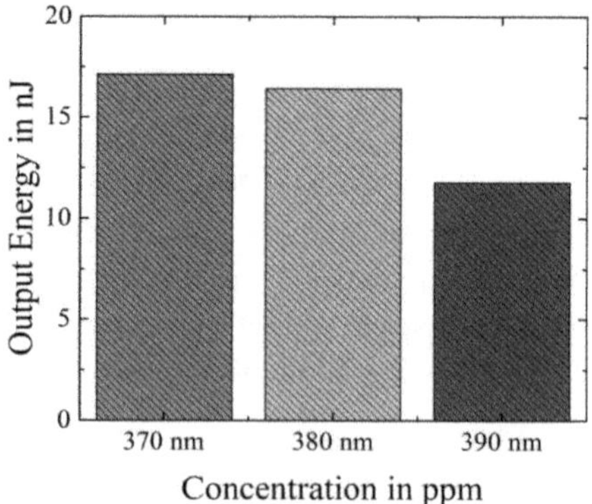

Figure 5.42: Output energy of the sample with different grating periods (370 nm, 380 nm, and 390 nm).

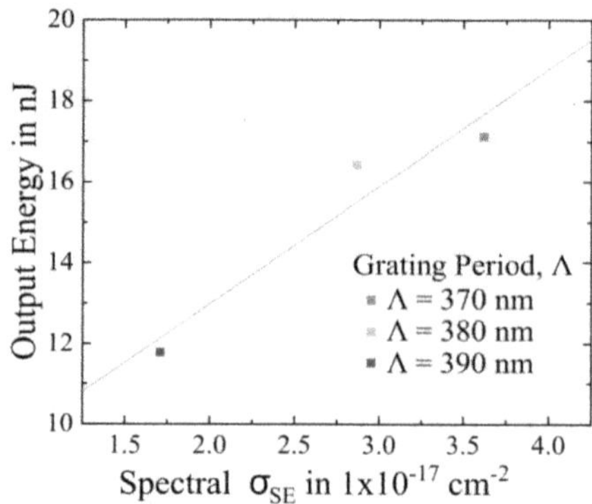

Figure 5.43: Relationship between the spectral stimulated emission cross section and the output energy under the influence of different grating periods.

5.5.3 Dye Variation

The results shown in the chapter before are the outcome of the sample doped with Rh6G. Changes of dopants in the gain medium affect the lasing properties, which is also a feature of a laser, on top of narrowing of FWHM and a clear lasing threshold. In order to find out the feasibility of other dyes in PMMA, several dyes were tested in this research. Subsequently, there were only six dyes, which could function well together with PMMA, which were Rh6G, RhB, LumO, P597, DCJTB, and DCM2. The concentration of the dyes was specified at 400 ppm according to the result shown in Chapter 5.2. The applied grating period was based on the result in Table 5.2. By using different dyes, the lasing wavelength could be tuned, in which the lasing properties would be altered. Therefore, the characteristics like lasing spectrum, lasing threshold, and PER were investigated. Moreover, the output energy, the sample lifetime, the pulse repetition rate, and the pulse duration of the samples were also determined. Most of the results of Rh6G had been shown in the chapter before, but it was represented here again as a reference.

Lasing Spectrum of Dye Variation

First of all, the lasing spectrum of the samples under the influence of different dyes was examined. The outcome is displayed in Figure 5.44, in which the samples were excited before the lasing threshold (0.10 µJ) and after the lasing threshold (1.69 µJ). As a reference, the lasing spectrum of Rh6G had a peak wavelength of 575.05 nm and a FWHM of 0.83 nm. For the lasing spectrum of RhB, the peak wavelength was detected at 593.28 nm and the FWHM was 0.81 nm. By using LumO, the lasing spectrum consisted of a peak wavelength of 579.74 nm and a FWHM of 0.70 nm. With the application of P597, the peak wavelength of the lasing spectrum was recognised at 572.45 nm and the FWHM was 0.57 nm. When DCJTB was employed, the peak wavelength of the lasing spectrum shifted to 608.91 nm, while the FWHM was 0.90 nm. When a DCM2 doped sample was measured, it was found that the peak wavelength of the lasing spectrum was 606.30 nm and the FWHM implied 0.93 nm. In short, by using these six various dyes, the peak wavelength could be tuned at the range from 572 nm to 609 nm, approximately, which had a wavelength shift of about 37 nm. Furthermore, all the detected FWHM was under 1 nm. It was observed that the FWHM was influenced by the spectral modal gain. The relationship between the FWHM and the spectral optical gain could be explained by using Equation 3.11. The spectral optical gain according to the peak wavelength was taken according to the Figure 5.15. The sample doped with P597 had the highest value of spectral optical gain (68.70 cm^{-1}) among the samples. Second, the spectral optical gain of LumO doped sample was evaluated for a value of 55.18 cm^{-1}. The third highest spectral optical gain belonged to RhB doped sample, which was 47.91 cm^{-1}. It was followed by the Rh6G doped sample, which had a spectral optical gain of 44.94 cm^{-1}. The spectral optical gains of DCJTB and DCM2 were relatively small, which were 32.93 cm^{-1} and 32.34 cm^{-1}. The linear relationship between the FWHM and the spectral optical gain is illustrated in Figure 5.45. The FWHM became smaller with increasing spectral modal gain.

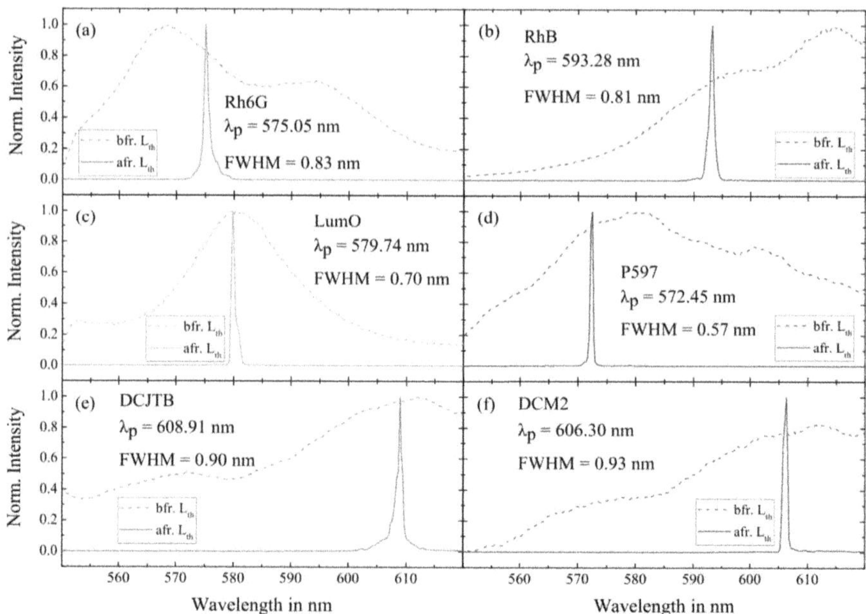

Figure 5.44: Lasing spectra of the sample doped with different dyes: (a) Rh6G, (b) RhB, (c) LumO, (d) P597, (e) DCJTB, and (f) DCM2.

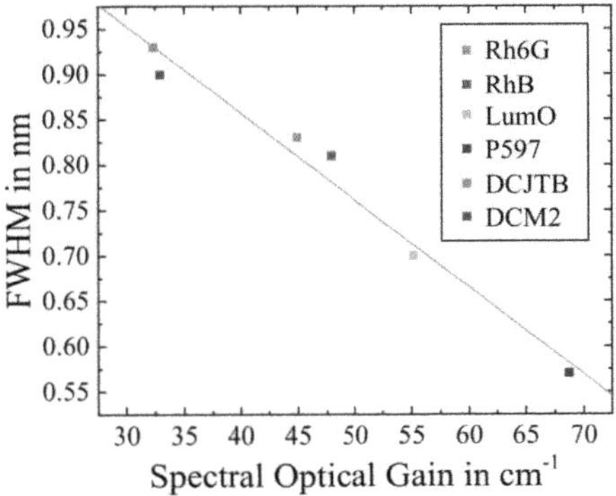

Figure 5.45: Relationship between the spectral optical gain and the FWHM of the respective spectra.

Lasing Threshold of Dye Variation

On top of the lasing spectrum, the lasing threshold of the sample with different dyes was also determined, which is presented in Figure 5.46. The lasing threshold of the Rh6G doped sample was 0.311 µJ. Meanwhile, the lasing threshold of the RhB doped sample was higher than the Rh6G doped sample, which was 0.775 µJ. By using the sample doped with LumO, a lasing threshold of 0.454 µJ was tested. Among the samples, the P597 doped sample had the lowest lasing threshold, which was 0.253 µJ. On the other hand, the lasing threshold of the sample doped with DCJTB had the highest lasing threshold, which was 1.254 µJ. When the DCM2 doped sample was applied, a lasing threshold of 1.227 µJ was detected. According to Equation 3.7, the lasing threshold could be affected by the spectral stimulated emission cross section. The spectral stimulated emission cross section respective to the laser dye was calculated by using the spectral modal gain according to Equation 3.4. The relationship between the spectral emission cross section and the lasing threshold was shown in Figure 5.47. With increasing spectral stimulated emission cross section, the lasing threshold decreased. It was observed that saturation was reached, when the P597 doped sample was used, which had the highest stimulated emission cross section. The lasing threshold could not be further decreased, which was due to the absorption property of the matrix material, PMMA. Besides, the dyes possibly have different excited absorption transition states, which caused this result.

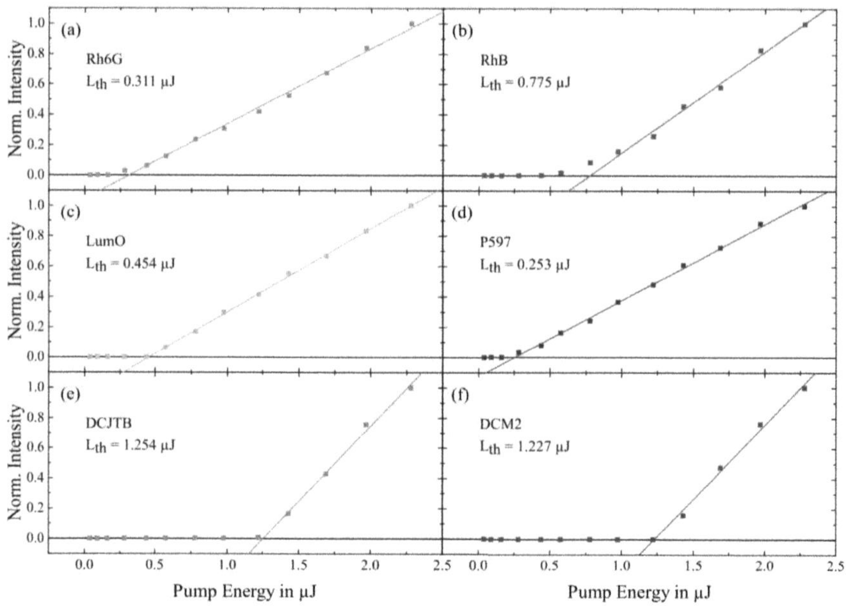

Figure 5.46: Lasing threshold of the sample doped with different dyes: (a) Rh6G, (b) RhB, (c) LumO, (d) P597, (e) DCJTB, and (f) DCM2.

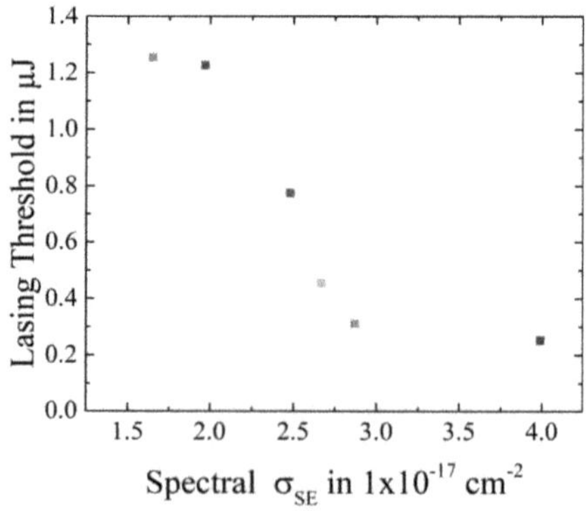

Figure 5.47: Relationship between the spectral stimulated emission cross section and the lasing threshold of the sample doped with different dyes.

PER of Dye Variation

On top of that, the PER of the samples from the impact of different dyes as dopants was also evaluated at a pump energy of 1.69 μJ, which is presented in Figure 5.48. The PERs of the sample doped with Rh6G, RhB and LumO was nearly identical, which were 30.88 dB, 30.65 dB, and 30.89 dB, respectively. The PER of the P597 doped sample was the highest among the samples, which was 31.12 dB. On the other side, the PERs of DCJTB and DCM2 doped samples were considered low, which were 28.06 dB and 28.01 dB, accordingly. The PER was not really affected by the laser dyes, which was in the range of 28 dB to 30 dB, approximately. Thus, the sample had a PER of > 28 dB, once it started lasing.

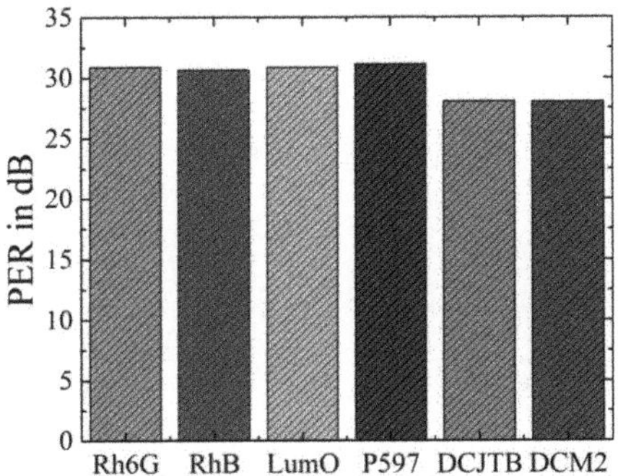

Figure 5.48: PER of the sample doped with different dyes (Rh6G, RhB, LumO, P597, DCJTB, and DCM2).

Output Energy of Dye Variation

Figure 5.49 shows the output energy of the sample doped with different dyes by using a pump energy of 1.69 μJ. Among the samples, the P597 doped sample had the highest output energy, which was 18.04 nJ. The second-highest output energy belonged to the sample doped with Rh6G, which had an output energy of 16.43 nJ. Then, it was followed by the LumO doped sample, in which an output energy of 14.01 nJ was determined. Next, the sample doped with RhB consisted of an output energy of 12.42 nJ. Contrary to this, DCJTB and DCM2 doped samples had relatively low output energies, which were 5.80 nJ and 6.18 nJ, correspondingly. The trend of output energy could be explained by using the spectral stimulated emission cross section, which indicated the emission efficiency. The trend between the stimulated emission cross section and the output energy was displayed in Figure 5.50. It was observed that the output energy increased when the spectral stimulated cross section was higher. Saturation was reached by the P597 doped sample, which might due to the absorption of the matrix material, PMMA. The applied dyes might have different excited absorption transition states, so no obvious trend could be declared.

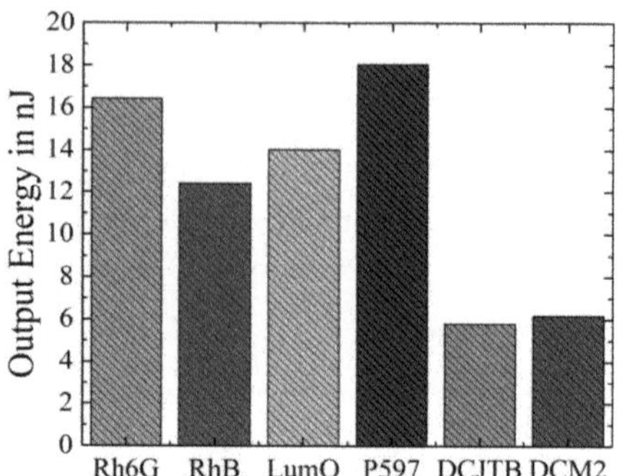

Figure 5.49: Output energy of the sample doped with different dyes (Rh6G, RhB, LumO, P597, DCJTB, and DCM2).

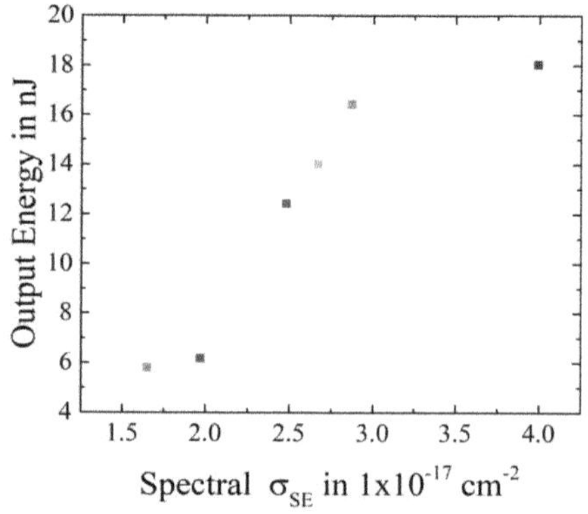

Figure 5.50: Relationship between the spectral stimulated emission cross section and the output energy under the influence of different dyes.

Sample Lifetime of Dye Variation

In addition to the lasing properties mentioned above, the sample lifetime of samples doped with different dyes was also evaluated. It was measured in the way that how many pump pulses of the excitation source the sample could withstand, in which the output intensity dropped to 50 % of its maximum. In such a way, pump pulses were selected as the unit of the sample lifetime. For the measurement, the used pump energy was fixed at 1.69 µJ. The outcome was presented in Figure 5.51. Among the samples, the P597 doped sample had the longest sample lifetime, which was 33 000 pump pulses. Next, the LumO doped sample had a sample lifetime of 23 000, which was the second-highest. Then, the third-place belonged to the RhB doped sample, which was 20 000 pump pulses. Afterwards, it was followed by the sample doped with Rh6G, in which a sample lifetime of 16 000 pump pulses was examined. Conversely, the sample lifetimes of the DCJTB and the DCM2 doped samples were considered low, which were 2500 pump pulses and 3000 pump pulses, respectively. It was discovered that the sample lifetime was in a linear relationship to the fluorescence time. Figure 5.52 provides the information that the sample lifetime increased when the fluorescence lifetime was longer.

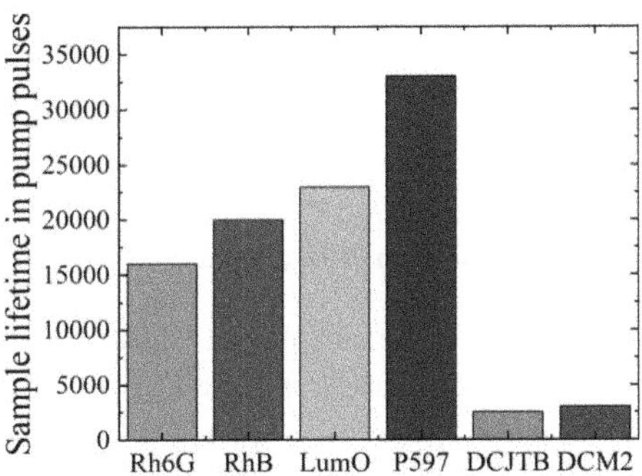

Figure 5.51: Sample lifetime of the sample doped with different dyes (Rh6G, RhB, LumO, P597, DCJTB and DCM2).

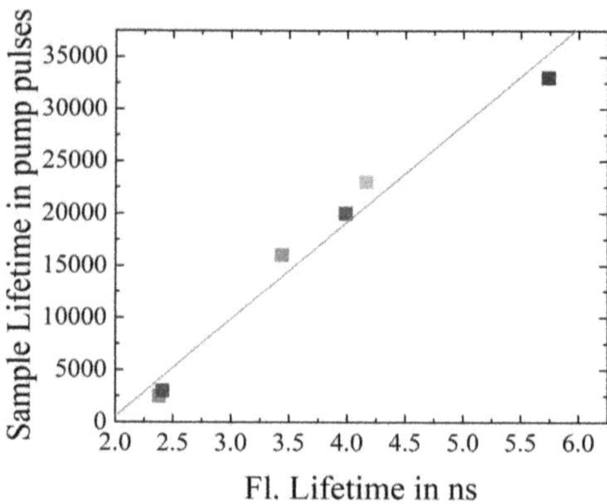

Figure 5.52: Relationship between the fluorescence lifetime and the sample lifetime under the influence of different dyes.

Pulse Repetition Rate and Pulse Duration of Dye Variation

Moreover, it should not be neglected to measure the pulse repetition rate and the pulse duration of the sample doped with various dyes to check if the dye variation had an influence on it. An overview of the result is illustrated in Table 5.4. The excitation source itself had a pulse repetition rate of 100 Hz and a pulse duration of 570±20ps, according to the data provided by the production firm. The samples were excited by a pump energy of 1.69 μJ. The pulse repetition rate of all the samples was the same as the excitation source, which was 100 Hz. Regarding the measurement of pulse duration, the sample doped with Rh6G had a value of 577 ps. By using the sample doped with RhB, the pulse duration was 576 ps. The pulse duration became 568 ps, when the LumO doped sample was applied. With the employment of P597 doped samples, a pulse duration of 572 ps was detected. Meanwhile, the pulse duration of the DCJTB doped sample was 572 ps. When the sample was doped with DCM2, a pulse duration of 578 ps was evaluated. As an outline, all the measured data were agreed with the data of pulse duration provided by the production firm, which was 570±20ps. It could be deduced that the dye variation does not affect the pulse repetition rate and the pulse duration. By assuming the laser pulse was an ideal rectangular shape, the maximum power of the sample could be computed, in which the output energy was divided by the pulse duration (570 ps). Among the samples, the highest maximum power was determined by using the P597 doped sample, which was 31.65 W. Next, the Rh6G doped sample contained a maximum power of 28.82 W. The third-place belonged to the LumO doped sample, which consisted of a maximum power of 24.58 W. Then, it was followed by the RhB doped sample with a maximum power of 21.79 W. The DCJTB and the DCM2 doped samples had relatively low maximum powers, which were 10.18 W and 10.84 W, correspondingly.

Table 5.4: The pulse repetition rate and the pulse duration of the samples doped with different dyes.

Sample	Pulse repetition rate /Hz	Pulse duration /ps	Max. Power /W
Excitation source	100	570±20	-
Rh6G	100	577	28.82
RhB	100	576	21.79
LumO	100	568	24.58
P597	100	572	31.65
DCJTB	100	572	10.18
DCM2	100	578	10.84

5.5.4 Application Potential

According to the result shown, D5 was the best design among the samples. By comparing the samples using different dyes as dopants, P597 presented the best outcomes. It had the smallest FWHM (0.57 nm) with a peak wavelength of approximately 572 nm. In addition, the lasing threshold (0.253 µJ) was the smallest. Besides, the output energy (18.04 nJ) was the highest. Moreover, the sample lifetime (33 000 pump pulses) was the longest. Therefore, P597 doped PMMA with the D5 was selected for a further test of application.

The capability of refractive index sensing of a DFB laser was researched by Morales-Vidal et al. [14]. In their work, the refractive index of the tested solution was manipulated by mixing water and glycerin. By modifying the idea as a refractive index sensor, the sample could be used to measure the concentration of sugar solution. This is due to the fact that the refractive index of sugar solution rises with increasing concentration. In such a way, a biosensor might be generated based on a DFB laser.

According to [148], the refractive index of the sugar concentration (n_{sugar}) can be calculated using the equation below:

$$n_{sugar} = n_w - aC. \tag{5.1}$$

where, n_w is the refractive index of water at 25 °C (n_w = 1.3324 at 589 nm), a denotes a specific increment of a refraction (0.001 43) and C is sugar concentration in g 100 mL or wt %.

To prepare solutions with different sugar concentrations, various weight percentages of glucose were dissolved in water. In this experiment, the determination of the refractive index of the sugar solution was accomplished by using the refractometric method. For this purpose, an Atago refractometer with 589 nm as incident light was employed. Figure 5.53 illustrates the experimental values and the theoretical values of the refractive index of sugar concentration. When the sugar concentration raised, the refractive index increased. It was in a linear relationship. Therefore, it could be assumed that the refractive index was associated with the sugar percentage. Besides, it is noticed that there was a good agreement between the theoretical and experimental values.

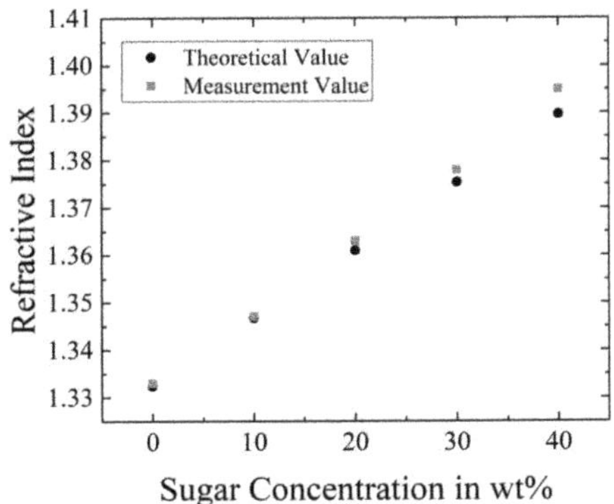

Figure 5.53: The refractive index of sugar solution became higher with increasing sugar concentration.

As mentioned above, the P597 doped sample was selected for this experiment. The sample was modified by attaching a cover lid on the sample (see Figure 5.54). Different concentrations of sugar solutions were injected into the modified sample design. Then, the sample was put into the laser setup in Figure 4.18 to determine the lasing spectrum with a pump energy of 1.69 µJ. The result is displayed in Figure 5.55. The lasing spectrum of P597 without any modification was shown as a reference, which had a peak wavelength of 572.48 nm. When only water was inserted (0wt% of sugar solution), the main peak wavelength was slightly shifted to a longer wavelength, which was seen at 572.60 nm. By using the solution with a 10wt% sugar concentration, the main peak wavelength was detected at 572.78 nm. By further increasing the sugar concentration to 20wt%, the peak wavelength was found at 573.38 nm. When the sample had a sugar concentration of 30wt%, a peak wavelength of 574.59 nm was measured. The sample with a sugar concentration of 40wt% consisted of a peak wavelength of 576.22 nm. With the implementation of all the sugar solutions with different concentrations, two small peaks were determined at around 570.06 nm and 576.65 nm. These two peak wavelengths demonstrated the occurrence of water because they were already detected when the sample with only water (0wt% of sugar solution) was applied. More peak wavelengths showed up in the samples when the applied sugar concentrations were higher (30wt% and 40wt%). This could be due to the reason that crystallisation of sugar occurred with higher sugar concentrations.

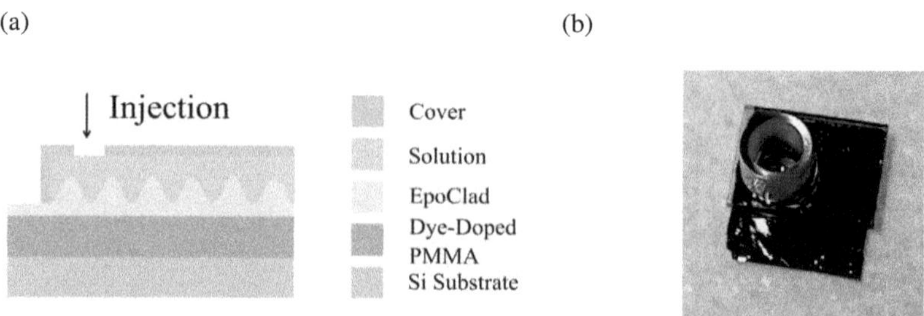

Figure 5.54: (a) Cover lid was attached on the sample with the structure of D5. Solution with different sugar concentrations was added to the sample to check its influence on the optical property, while the actual sample was shown in (b).

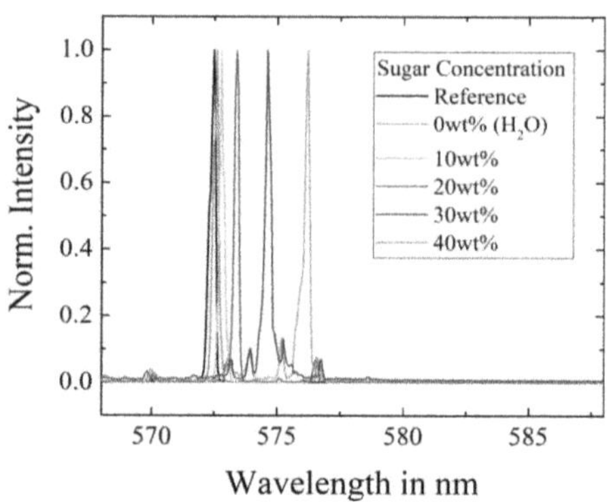

Figure 5.55: Lasing spectra of P597 doped samples in contact with solution of different sugar concentrations.

The difference of the main peak wavelength of various sugar concentrations to the peak wavelength of the reference was computed. The relationship of the wavelength shift to the refractive index of different sugar concentrations is illustrated in Figure 5.56. Apparently, the main peak wavelength moved to a longer wavelength exponentially with increasing sugar concentration. The result demonstrates that the sample in this work had a potential for sensing applications. Further research should be carried on for fine detection.

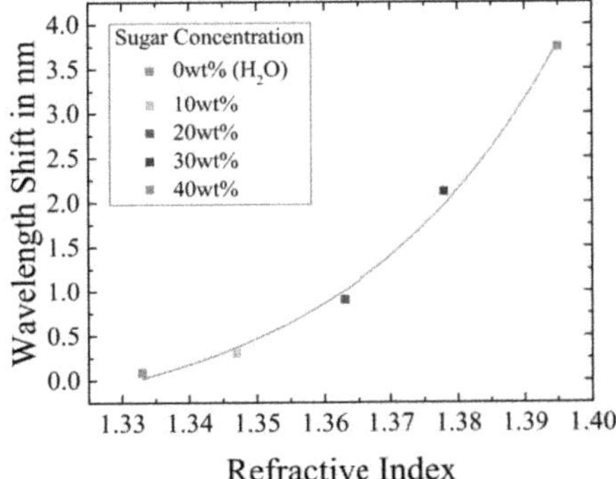

Figure 5.56: Relationship between the refractive index and the wavelength shift with respective to the reference under the effect of different sugar concentrations.

6 Conclusion

In this research, the feasibility of organic dye doped PMMA DFB laser has been tested. Several sample designs have been introduced to determine the best sample structure. By using the best sample architecture, the thickness of the resonator and the active layer have been manipulated. The dye dopants concentration in the gain material has been varried to check its effect on lasing properties. Besides, various grating periods have been employed, which functioned as the resonator of DFB laser. In addition, the functionality of several laser dyes doped in PMMA as DFB laser has also been investigated. Moreover, an application has been discovered to detect solution with different sugar concentrations.

To determine the sample design with the best performance, six different sample models have been proposed. In this test, Rh6G has been doped with PMMA as matrix and the Rh6G concentration has been fixed at 400 ppm. Within these six designs, various structures have been introduced, such as merely thin-film, direct grating on gain material, thin-film covered with EpoClad, direct grating on gain material covered with EpoClad, gain medium sandwiched between substrate and EpoClad layer with a grating structure on top of it, and gain medium on top of a grating structure made up of EpoClad layer. It has been discovered that the sample with the gain medium covered with EpoClad with grating structure (D5) presented the best result. The sample model with the best outcome has been applied for further experiments.

Besides, the thickness of the dye-doped PMMA layer and the EpoClad layer have been manipulated to check the improvement of lasing performance. The thickness of EpoClad layer has been varied using dilution with Gbl and speed of spin-coating. The output energy worsened when the dilution was too high. The organisation of grating in EpoClad was limited by the speed of spin-coating. On top of this, the thickness of PMMA layer has been controlled by speed of spin-coating. It was no longer single-mode, when the thickness was too high. Besides, the lasing properties aggravated with decreasing thickness of PMMA layer. As the best sample structure, the thickness of EpoClad layer has been fixed at 661 nm, while the thickness of PMMA layer has been kept at 1200 nm.

To prove that it is a laser, the active layer and the resonator have been regulated because the lasing properties could be strongly influenced. At first, the dye concentration of dopants has been varied, which was Rh6G in this case. When the concentration (100 ppm) was too low, a relatively big FWHM was measured and no obvious lasing threshold was observed because only ASE was achieved. There was a small shift of peak wavelength when the concentration was increased from 200 ppm to 400 ppm. With increasing concentration, the FWHM, the lasing threshold, and the output energy were also improved. For this reason, the concentration has been maintained at 400 ppm for additional investigations.

In the next step, the grating period in the grating structure has been adjusted, which worked as a resonator for the DFB laser. 400 ppm Rh6G concentration was used as the dopant in this experiment. There were in total three different grating periods applied: 370 nm, 380 nm, and 390 nm. There was in total a wavelength-shift of 25 nm by using these three grating periods, while the FWHM was less than 1 nm. When the spectral optical gain increased, the FWHM became smaller. It is observed that the lasing threshold decreased when the grating period was smaller. This can be explained by using the stimulated emission cross section. By using a smaller grating period, the peak wavelength occurred at a wavelength that corresponded to a higher optical gain. The higher the optical gain, the higher the stimulated emission cross section. Therefore, it could be declared that the lasing threshold minimised with increasing stimulated emission cross section. Moreover, the output energy increased as well, when the stimulated emission cross section raised.

Over and above, it is interesting to examine the workability of other laser dyes with PMMA as DFB laser. It has been noticed that there have been six well-functioning laser dyes, which were Rh6G, RhB, LumO, P597, DCJTB and DCM2. For the measurement of lasing properties, the grating period used was based on the peak of the photoluminescence spectra of respective laser dyes. It has been found out that there was about 37 nm of wavelength shift by using these laser dyes. At the meantime, a FWHM of less than 1 nm was detected. The FWHM decreased with increasing optical gain. Furthermore, the lasing threshold became less with a higher stimulated emission cross section. Besides, when the stimulated emission cross section increased, the output energy was also enhanced. In addition, when the sample had a longer fluorescence lifetime, the sample lifetime tended to become longer. On top of these, it has been discovered that the PER of the samples was > 28 dB, once the sample reach the lasing state. The best performance was shown by the P597 doped sample. By using the P597 doped sample, it has been recognised that there is potential in detecting sugar concentration. For the range from 0wt% to 40wt% sugar concentration, there was an approximate 4 nm of red-shift.

In a conclusion, the feasibility of organic dye-doped PMMA DFB thin-film laser has been practised. The sample with the design, in which the active layer has been covered between the substrate and the EpoClad layer assembled with grating structure, has shown the best outcome. Besides, the thickness of the gain medium layer and the grating structure layer have been manipulated to assure the lasing performance. Dye concentration, resonator, and selective dyes can influence the lasing outcome, which can be explained by the material characteristics. The potential in sensing application has been discovered. The functionality of organic dye-doped DFB laser using PMMA as a matrix has been proven. Further investigations should be carried on for more applications.

Acknowledgements

First, I would like to express my appreciation to Prof. Dr. Wolfgang Kowalsky for giving me a chance to work under his group and providing me an opportunity to work in this interesting research field. Second, the kindness of Prof. Dr. Bernhard Wilhelm Roth is also been acknowledged for becoming my second supervisor of this work. Besides, thank is given to Prof. Dr. -Ing Jörg Schöbel as my chairman for my Phd colloquium. Moreover, my gratitude is also given to Dr. Hans-Hermann Johannes, who is my group leader. With his provided working atmosphere and assistance, I can only succeed in my work. Next, I would like to thank you the following fundings: LaPOF research network (EFRE-SER 85003655), Cluster of Excellence PhoenixD (Exc 2122, Project ID 390833453) and DFG-project PolySens (AOBJ: 654666).

The quality of work was improved with correction, discussion, and support from many people. Therefore, I would like to take this chance to give my thanks to the following people:

- Marko Cehovski for scientific discussion.
- Simone Schulze from ICTV TU Braunschweig for SEM measurements.
- Dinara Samigullina from IAPP TU Dreden for AFM measurements.
- Christian Hänisch from IAPP TU Dresden for refractive index measurements.
- Stephanie Michel from IPAT TU Braunschweig for the measurement of grating structures.
- all the working colleagues for working environment and support of measurement instruments.
- Patrick Bethke for scientific discussion and motivation.
- Frederike Lompa for sample preparation.
- Jana Kielhorn for chemistry discussion and encouragement.

- Henrik-Alexander Christ and Fuzhao Li for scientific discussion.
- Secretary and admin, who are Carola Baaske, Christa Vögel, Petra Groß-Ulrich, and Sabrina Klima for the explanation of bureaucracy and encouragement.
- Kathleen Möhring for technical support.
- the "werkstatt" group: Olaf Flechtner, Anne Feichert and Frank Denecke for manufacture of components for my measurement setups.
- Katja Rudnik, Yu Tokura, Niklas Henning, Franz Selzer and Katrin Ortstein for grammar corrections and improvement suggestions.

All the nice events are appreciated, such as Christmas celebration, BBQ-lunch, Frankfurt day trip, 111 birthday celebration and etc. I am very grateful to my friends (Thies Rudnik, Hartz VI group, Rotpot Team and Wednesday cooking group) and family (my mum, Cheah Sooi Nya and my sister Ang Yew See) for helping me cope with all the stress. There are a lot of supports from many people that I have taken. Without their help, I do not think that I would have been able to complete my dissertation.

Abbreviations

α_{abs} absorption coefficient. 57, 58, 62, 65

AFM atomic force microscope. 44, 54, 56

ASE amplified stimulated emission. 22, 23, 45, 50, 71, 86, 89, 113

DCJTB 4-(Dicyanomethylene) -2-tert-butyl-6- (1,1,7,7-tetramethyljulolidin-4-yl-vinyl)- 4H-pyran. iii, 35, 62, 64–66, 68, 99–107, 113

DCM2 4-(Dicyanomethylene)-2-methyl-6-julolidyl-9-enyl-4H- pyran. iii, 35, 62, 64–66, 68, 99–107, 113

DFB distributed feedback. 25, 26, 28–31, 37, 108, 112–114

FWHM Full Width at Half Maximum. 3, 23, 29, 50, 52, 58, 62, 65, 71, 72, 74, 77, 81–83, 85–90, 94, 95, 99–101, 113

Gbl gamma-Butyrolactone. 34, 39, 54, 78, 112

HOMO highest occupied molecular orbital. 12

laser light amplification by stimulated emission of radiation. 16, 17, 37, 99, 112–114

LCAO linear combination of atomic orbitals. 11

LUMO lowest unoccupied molecular orbital. 12

LumO Lumogen Orange. iii, 35, 62, 63, 65–67, 99–107, 113

OLEDs organic light-emitting diodes. 1, 30

OSLs organic semiconductor lasers. 1, 2

OSSLs organic solid-state lasers. 1, 2, 18, 29

P597 Pyrromethene 597. iii, iv, 35, 62, 64–66, 68, 99–110, 113

PDMS polydimethylsiloxane. 37, 38, 40–42

PER polarisation extinction ratio. 50, 51, 75–77, 89, 91, 92, 97, 99, 103, 113

PLQY photoluminescence quantum yield. 2, 14, 46, 58, 59, 65

PMMA poly(methyl methacrylate). 1–3, 16, 32, 33, 35, 37–39, 42, 45, 46, 53–57, 60, 62, 69–74, 76–78, 84–89, 99, 101, 104, 108, 112–114

PPV poly(p-phenylenevinylene). 1

PS polystyrene. 2, 32

Rh6G Rhodamine 6G. iii, 1, 29, 35, 39, 57–67, 69, 77, 84, 85, 89–94, 99–107, 112, 113

RhB Rhodamine B. iii, 29, 35, 62–67, 99–107, 113

rpm rounds per minute. 36

SEM scanning electron microscope. 44, 53, 55, 56

Si silicon. 39, 47, 52

TE transverse electrical. 7, 8, 75

TM transverse magnetic. 7, 8, 75

VSL variable stripe length. 47–50, 59, 66

List of Figures

Bibliography

[1] H. Uoyama et al. "Highly efficient organic light-emitting diodes from delayed fluorescence". In: *Nature* 492.7428 (2012), pp. 234–238.

[2] Y. Tao et al. "Achieving optimal self-adaptivity for dynamic tuning of organic semiconductors through resonance engineering". In: *Journal of the American Chemical Society* 138.30 (2016), pp. 9655–9662.

[3] P. Y. Ang et al. "Inside or outside: Evaluation of the efficiency enhancement of OLEDs with applied external scattering layers". In: *Scientific reports* 9.1 (2019), pp. 1–10.

[4] Y. Liu et al. "Aggregation and morphology control enables multiple cases of high-efficiency polymer solar cells". In: *Nature communications* 5.1 (2014), pp. 1–8.

[5] Z. He et al. "Enhanced power-conversion efficiency in polymer solar cells using an inverted device structure". In: *Nature photonics* 6.9 (2012), pp. 591–595.

[6] G. Giri et al. "Tuning charge transport in solution-sheared organic semiconductors using lattice strain". In: *Nature* 480.7378 (2011), pp. 504–508.

[7] H. E. Katz, Z. Bao, and S. L. Gilat. "Synthetic chemistry for ultrapure, processable, and high-mobility organic transistor semiconductors". In: *Accounts of Chemical Research* 34.5 (2001), pp. 359–369.

[8] B. K. Yap et al. "Simultaneous optimization of charge-carrier mobility and optical gain in semiconducting polymer films". In: *Nature Materials* 7.5 (2008), pp. 376–380.

[9] Y. Jiang et al. "Organic solid-state lasers: a materials view and future development". In: *Chemical Society Reviews* 49.16 (2020), pp. 5885–5944.

[10] C. Vannahme et al. "Plastic lab-on-a-chip for fluorescence excitation with integrated organic semiconductor lasers". In: *Optics express* 19.9 (2011), pp. 8179–8186.

[11] Y. Oki et al. "Multiwavelength distributed-feedback dye laser array and its application to spectroscopy". In: *Optics letters* 27.14 (2002), pp. 1220–1222.

[12] T Woggon, S Klinkhammer, and U Lemmer. "Compact spectroscopy system based on tunable organic semiconductor lasers". In: *Applied Physics B* 99.1-2 (2010), pp. 47–51.

[13] J. Clark and G. Lanzani. "Organic photonics for communications". In: *Nature photonics* 4.7 (2010), pp. 438–446.

[14] M. Morales-Vidal et al. "Distributed feedback lasers based on perylenediimide dyes for label-free refractive index sensing". In: *Sensors and Actuators B: Chemical* 220 (2015), pp. 1368–1375.

[15] Y. Wang et al. "LED pumped polymer laser sensor for explosives". In: *Laser & photonics reviews* 7.6 (2013), pp. L71–L76.

[16] B. Soffer and B. McFarland. "Continuously tunable, narrow-band organic dye lasers". In: *Applied physics letters* 10.10 (1967), pp. 266–267.

[17] N Karl. "Laser emission from an organic molecular crystal". In: *physica status solidi (a)* 13.2 (1972), pp. 651–655.

[18] C. W. Tang and S. A. VanSlyke. "Organic electroluminescent diodes". In: *Applied physics letters* 51.12 (1987), pp. 913–915.

[19] F. Hide et al. "Semiconducting polymers: a new class of solid-state laser materials". In: *Science* 273.5283 (1996), pp. 1833–1836.

[20] N Tessler, G. Denton, and R. Friend. "Lasing from conjugated-polymer microcavities". In: *Nature* 382.6593 (1996), pp. 695–697.

[21] Y Yang. "G. a. Turnbull, and IDW Samuel". In: *Appl. Phys. Lett* 92 (2008), p. 163306.

[22] W. Lai, R Xia, and Q. He. "P. a. Levermore, W. Huang and DDC Bradley". In: *Adv. Mater* 21 (2009), pp. 355–360.

[23] M. C. Gather and S. H. Yun. "Single-cell biological lasers". In: *Nature Photonics* 5.7 (2011), pp. 406–410.

[24] A. S. Sandanayaka et al. "Toward continuous-wave operation of organic semiconductor lasers". In: *Science advances* 3.4 (2017), e1602570.

[25] D.-H. Kim et al. "High-efficiency electroluminescence and amplified spontaneous emission from a thermally activated delayed fluorescent near-infrared emitter". In: *Nature Photonics* 12.2 (2018), pp. 98–104.

[26] A. S. Sandanayaka et al. "Indication of current-injection lasing from an organic semiconductor". In: *Applied Physics Express* 12.6 (2019), p. 061010.

[27] F. P. Schäfer, W. Schmidt, and J. Volze. "Organic dye solution laser". In: *Applied Physics Letters* 9.8 (1966), pp. 306–309.

[28] I. D. W. Samuel and G. A. Turnbull. "Organic semiconductor lasers". In: *Chemical reviews* 107.4 (2007), pp. 1272–1295.

[29] P. G. Boj et al. "Organic distributed feedback laser to monitor solvent extraction upon thermal annealing in solution-processed polymer films". In: *Sensors and Actuators B: Chemical* 232 (2016), pp. 605–610.

[30] J. A. Quintana et al. "An Efficient and Color-Tunable Solution-Processed Organic Thin-Film Laser with a Polymeric Top-Layer Resonator". In: *Advanced Optical Materials* 5.19 (2017), p. 1700238.

[31] V. Bonal et al. "Solution-processed nanographene distributed feedback lasers". In: *Nature communications* 10.1 (2019), pp. 1–10.

[32] V. Bonal et al. "Controlling the emission properties of solution-processed organic distributed feedback lasers through resonator design". In: *Scientific reports* 9.1 (2019), pp. 1–10.

[33] S. Yuyama et al. "Solid state organic laser emission at 970 nm from dye-doped fluorinated-polyimide planar waveguides". In: *Applied Physics Letters* 93.2 (2008), p. 252.

[34] O. Mhibik et al. "Broadly tunable (440–670 nm) solid-state organic laser with disposable capsules". In: *Applied Physics Letters* 102.4 (2013), p. 041112.

[35] *Leitfaden "Sichtbare und infrarote Strahlung"*. URL: http://www.fs-ev.org/fileadmin/user_upload/04_Arbeitsgruppen/08_Nichtionisierende_Strahlung/02_Dokumente/Leitfaeden/Leitfaden-SB-IR-AKNIR-15122011_b.pdf (visited on 07/03/2020).

[36] *The electromagnetic spectrum*. URL: https://chem.libretexts.org/Bookshelves/Introductory_Chemistry/Map%3A_Introductory_Chemistry_(Tro)/09%3A_Electrons_in_Atoms_and_the_Periodic_Table/9.03%3A_The_Electromagnetic_Spectrum (visited on 03/12/2020).

[37] M. Born and E. Wolf. *Principles of optics: electromagnetic theory of propagation, interference and diffraction of light*. Elsevier, 2013.

[38] G. Lu. "Organic semiconductor". In: *University of Rochester, Rochester, NY* (2006), pp. 1–4.

[39] T Benstem. "Lumineszenz-Dynamik und stimulierte Emission von organischen Dünnschichten". Dissertation. TU Braunschweig, 2002.

[40] F. Goos and H Hänchen. "Über das Eindringen des totalreflektierten Lichtes in das dünnere Medium". In: *Annalen der Physik* 435.5 (1943), pp. 383–392.

[41] H. K. Lotsch. "Reflection and refraction of a beam of light at a plane interface". In: *JOSA* 58.4 (1968), pp. 551–561.

[42] W. Bludau. *Lichtwellenleiter in Sensorik und optischer Nachrichtentechnik*. Springer-Verlag, 2013.

[43] W. Glaser. *Photonik für Ingenieure: mit Aufgaben und Lösungen*. Verlag Technik, 1997.

[44] T Rabe. "Materialen und Bauelemententstrukturen für organische Laserdioden". Dissertation. TU Braunschweig, 2010.

[45] H. Fouckhardt. *Photonik*. Teubner StudienbÃ¼cher, 1994.

[46] W. Bludau. *Lichtwellenleiter in Sensorik und optischer Nachrichtentechnik*. Springer-Verlag, 2013.

[47] H. E. Katz, Z. Bao, and S. L. Gilat. "Synthetic chemistry for ultrapure, processable, and high-mobility organic transistor semiconductors". In: *Accounts of Chemical Research* 34.5 (2001), pp. 359–369.

[48] S. Reineke. "Organic Semiconductor lecture slides, lecture 2". In: (Summer Semester 2016).

[49] A. Kohler and H.Bassler. *Electronic Processes in Organic Semiconductors: An Introduction*.

[50] *Jablonski Diagram: Relaxation mechanism for excited state molecules*. URL: `http://www.shsu.edu/~chm_tgc/chemilumdir/JABLONSKI.html` (visited on 01/06/2021).

[51] Y. Xu and M. A. Schoonen. "The absolute energy positions of conduction and valence bands of selected semiconducting minerals". In: *American Mineralogist* 85.3-4 (2000), pp. 543–556.

[52] D. J. Gaspar and E. Polikarpov. *OLED fundamentals: materials, devices, and processing of organic light-emitting diodes*. CRC Press, 2015.

[53] F. B. Dias, T. J. Penfold, and A. P. Monkman. "Photophysics of thermally activated delayed fluorescence molecules". In: *Methods and Applications in Fluorescence* 5.1 (2017), p. 012001. URL: `http://stacks.iop.org/2050-6120/5/i=1/a=012001`.

[54] URL: `https://chem.libretexts.org/Textbook_Maps/Organic_Chemistry_Textbook_Maps/Map%3A_Organic_Chemistry_(Vollhardt_and_Schore)/15%3A_Benzene_and_Aromaticity%3A_Electrophilic_Aromatic_Substitution/15.03%3A_Pi_Molecular__Orbitals__of_Benzene` (visited on 07/12/2020).

[55] J. R. Lakowicz. *Principles of fluorescence spectroscopy*. Springer science & business media, 2013.

[56] J. B. Birks. *Organic molecular photophysics*. Vol. 2. Wiley-Interscience, 1973.

[57] S. Strickler and R. A. Berg. "Relationship between absorption intensity and fluorescence lifetime of molecules". In: *The Journal of chemical physics* 37.4 (1962), pp. 814–822.

[58] F. K. Kneubühl and M. W. Sigrist. *Laser*. Teubner, 1995.

[59] M. Müller-Wiegand. "Spontane Strukturbildung in Mehrschichtsystemen unter Verwendung Molekularer Gläser". PhD thesis. 2005.

[60] T. Rabe and T. Otto. "Skript zur Vorlesung Laser und Anwendungen". In: (2012).

[61] T Benstem. "Lumineszenz-Dynamik und stimulierte Emission von organischen Dünnschichten". In: *Cuvillier, Göttingen* (2002).

[62] P. Andrew et al. "Photonic band structure and emission characteristics of a metal-backed polymeric distributed feedback laser". In: *Applied Physics Letters* 81.6 (2002), pp. 954–956.

[63] M. Baldo, R. Holmes, and S. Forrest. "Prospects for electrically pumped organic lasers". In: *Physical Review B* 66.3 (2002), p. 035321.

[64] S. Chénais and S. Forget. "Recent advances in solid-state organic lasers". In: *Polymer International* 61.3 (2012), pp. 390–406.

[65] P. Bouguer. *Essai d'Optique sur la Gradation de la Lumière*. 1922.

[66] W. J. Miniscalco et al. "Measurement and analysis of cross sections for rare-earth-doped glasses". In: *Fiber Laser Sources and Amplifiers III*. Vol. 1581. International Society for Optics and Photonics. 1992, pp. 80–90.

[67] C Rulliere. *Femtesecond Laser Pulses, Principles and Experiments*. Springer, 1998.

[68] K. Shaklee and R. Leheny. "Direct determination of optical gain in semiconductor crystals". In: *Applied Physics Letters* 18.11 (1971), pp. 475–477.

[69] Y Sorek et al. "Light amplification in a dye-doped glass planar waveguide". In: *Applied physics letters* 66.10 (1995), pp. 1169–1171.

[70] S. Sze. *Semiconductor Devices*. John Wiley and Sons, 1985.

[71] L. W. Casperson. "Threshold characteristics of mirrorless lasers". In: *Journal of Applied Physics* 48.1 (1977), pp. 256–262.

[72] S. Van den Berg et al. "From amplified spontaneous emission to laser oscillation: dynamics in a short-cavity polymer laser". In: *Optics Letters* 24.24 (1999), pp. 1847–1849.

[73] A. Otomo et al. "Supernarrowing mirrorless laser emission in dendrimer-doped polymer waveguides". In: *Applied Physics Letters* 77.24 (2000), pp. 3881–3883.

[74] O Svelto. *Principles of Lasers*. Springer, 1998.

[75] D. Schneider. "Organischer Halbleiterlaser". Dissertation. TU Braunschweig, 2004.

[76] F. K. Kneubühl and M. W. Sigrist. *Laser*. Springer-Verlag, 2008.

[77] N. Ismail et al. "Fabry-Pérot resonator: spectral line shapes, generic and related Airy distributions, linewidths, finesses, and performance at low or frequency-dependent reflectivity". In: *Optics express* 24.15 (2016), pp. 16366–16389.

[78] H Zucker. "Optical resonators with variable reflectivity mirrors". In: *Bell System Technical Journal* 49.9 (1970), pp. 2349–2376.

[79] M. D. McGehee and A. J. Heeger. "Semiconducting (conjugated) polymers as materials for solid-state lasers". In: *Advanced Materials* 12.22 (2000), pp. 1655–1668.

[80] N Tessler, G. Denton, and R. Friend. "Lasing from conjugated-polymer microcavities". In: *Nature* 382.6593 (1996), pp. 695–697.

[81] M. Berggren et al. "Solid-state droplet laser made from an organic blend with a conjugated polymer emitter". In: *Advanced Materials* 9.12 (1997), pp. 968–971.

[82] G Ramos-Ortiz et al. "Temperature dependence of the threshold for laser emission in polymer microlasers". In: *Applied Physics Letters* 77.18 (2000), pp. 2783–2785.

[83] S. Frolov, Z. Vardeny, and K Yoshino. "Plastic microring lasers on fibers and wires". In: *Applied physics letters* 72.15 (1998), pp. 1802–1804.

[84] S. Frolov et al. "Ring microlasers from conducting polymers". In: *Physical review B* 56.8 (1997), R4363.

[85] I. D. W. Samuel and G. A. Turnbull. "Organic semiconductor lasers". In: *Chemical reviews* 107.4 (2007), pp. 1272–1295.

[86] F. K. Kneubühl and M. W. Sigrist. *Laser*. Springer-Verlag, 2008.

[87] H Kogelnik and C. Shank. "Coupled-wave theory of distributed feedback lasers". In: *Journal of applied physics* 43.5 (1972), pp. 2327–2335.

[88] R Kazarinov and C Henry. "Second-order distributed feedback lasers with mode selection provided by first-order radiation losses". In: *IEEE Journal of Quantum Electronics* 21.2 (1985), pp. 144–150.

[89] H Ghafouri-Shiraz and H Ghafouri-Shiraz. *Distributed feedback laser diodes and optical tunable filters*. Wiley Online Library, 2003.

[90] S Hansmann. "Spektrale Eigenschaften von Halbleiterlasern mit verteilter Rückkoplung". Dissertation. Technische Hochschule Darmstadt, 1994.

[91] K. J. Ebeling. *Integrierte Optoelektronik: Wellenleiteroptik. Photonik. Halbleiter*. Springer-Verlag, 2013.

[92] M. D. McGehee et al. "Amplified spontaneous emission from photopumped films of a conjugated polymer". In: *Physical Review B* 58.11 (1998), p. 7035.

[93] A. Deshpande and E. Namdas. “Lasing action of rhodamine B in polyacrylic acid films”. In: *Applied Physics B* 64.4 (1997), pp. 419–422.

[94] K. Parafiniuk et al. “Enlargement of the organic solid-state DFB laser wavelength tuning range by the use of two complementary luminescent dyes doped into the host matrix”. In: *Physical Chemistry Chemical Physics* 19.27 (2017), pp. 18068–18075.

[95] T Voss, D Scheel, and W Schade. “A microchip-laser-pumped DFB-polymer-dye laser”. In: *Applied Physics B* 73.2 (2001), pp. 105–109.

[96] A. Camposeo et al. “Electrically tunable organic distributed feedback lasers embedding nonlinear optical molecules”. In: *Advanced Materials* 24.35 (2012), OP221–OP225.

[97] M. G. Ramírez et al. “Distributed feedback lasers based on dichromated poly (vinyl alcohol) reusable surface-relief gratings”. In: *Optical Materials Express* 4.4 (2014), pp. 733–738.

[98] D. Nilsson, T. Nielsen, and A. Kristensen. “Solid state microcavity dye lasers fabricated by nanoimprint lithography”. In: *Review of Scientific instruments* 75.11 (2004), pp. 4481–4486.

[99] M. Gaal et al. “Imprinted conjugated polymer laser”. In: *Advanced Materials* 15.14 (2003), pp. 1165–1167.

[100] B. Wenger et al. “Mechanically tunable conjugated polymer distributed feedback lasers”. In: *Applied Physics Letters* 97.19 (2010), p. 237.

[101] E. B. Namdas et al. “Low thresholds in polymer lasers on conductive substrates by distributed feedback nanoimprinting: progress toward electrically pumped plastic lasers”. In: *Advanced materials* 21.7 (2009), pp. 799–802.

[102] S Klinkhammer et al. “A continuously tunable low-threshold organic semiconductor distributed feedback laser fabricated by rotating shadow mask evaporation”. In: *Applied Physics B* 97.4 (2009), p. 787.

[103] G. Tsiminis et al. “Nanoimprinted organic semiconductor laser pumped by a light-emitting diode”. In: *Advanced Materials* 25.20 (2013), pp. 2826–2830.

[104] M. Čehovski et al. “Single-Mode Polymer Ridge Waveguide Integration of Organic Thin-Film Laser”. In: *Applied Sciences* 10.8 (2020), p. 2805.

[105] S. Schauer et al. “Shape-memory polymers as flexible resonator substrates for continuously tunable organic DFB lasers”. In: *Optical Materials Express* 5.3 (2015), pp. 576–584.

[106] S. Klinkhammer et al. “Continuously tunable solution-processed organic semiconductor DFB lasers pumped by laser diode”. In: *Optics express* 20.6 (2012), pp. 6357–6364.

[107] C. Ge et al. “Large-area organic distributed feedback laser fabricated by nanoreplica molding and horizontal dipping”. In: *Optics Express* 18.12 (2010), pp. 12980–12991.

[108] J. Herrnsdorf et al. “Flexible blue-emitting encapsulated organic semiconductor DFB laser”. In: *Optics express* 18.25 (2010), pp. 25535–25545.

[109] R Xia et al. “Polyfluorene distributed feedback lasers operating in the green-yellow spectral region”. In: *Applied Physics Letters* 87.3 (2005), p. 031104.

[110] G. Heliotis et al. “Emission characteristics and performance comparison of polyfluorene lasers with one-and two-dimensional distributed feedback”. In: *Advanced Functional Materials* 14.1 (2004), pp. 91–97.

[111] X. Liu et al. “Organic semiconductor distributed feedback (DFB) laser as excitation source in Raman spectroscopy”. In: *Optics express* 21.23 (2013), pp. 28941–28947.

[112] K. Suzuki et al. “A continuously tunable organic solid-state laser based on a flexible distributed-feedback resonator”. In: *Japanese journal of applied physics* 42.3A (2003), p. L249.

[113] P. Görrn et al. “Elastically Tunable Self-Organized Organic Lasers”. In: *Advanced Materials* 23.7 (2011), pp. 869–872.

[114] S. Döring et al. “Electrically tunable polymer DFB laser”. In: *Advanced Materials* 23.37 (2011), pp. 4265–4269.

[115] J Wang et al. “A continuously tunable organic DFB laser”. In: *Microelectronic engineering* 78 (2005), pp. 364–368.

[116] R. Ozaki et al. “Tunable liquid crystal laser using distributed feedback cavity fabricated by nanoimprint lithography”. In: *Applied physics express* 1.1 (2008), p. 012003.

[117] S. Klinkhammer et al. “Voltage-controlled tuning of an organic semiconductor distributed feedback laser using liquid crystals”. In: *Applied physics letters* 99.2 (2011), p. 137.

[118] D. Moses. “High quantum efficiency luminescence from a conducting polymer in solution: A novel polymer laser dye”. In: *Applied Physics Letters* 60.26 (1992), pp. 3215–3216.

[119] L. Persano et al. “Monolithic polymer microcavity lasers with on-top evaporated dielectric mirrors”. In: *Applied physics letters* 88.12 (2006), p. 121110.

[120] T. Zhai, X. Zhang, and Z. Pang. “Polymer laser based on active waveguide grating structures”. In: *Optics express* 19.7 (2011), pp. 6487–6492.

[121] C. W. Tang and S. A. VanSlyke. “Organic electroluminescent diodes”. In: *Applied physics letters* 51.12 (1987), pp. 913–915.

[122] L. Duan et al. "Preparation of 8-hydroxyquinoline aluminum nanomaterials to enhance properties for green organic light-emitting diode devices". In: *Journal of the Society for Information Display* 29.6 (2021), pp. 466–475.

[123] *Polymers for a clear, transparent view.* URL: https://history.evonik.com/en/inventions/plexiglas (visited on 07/02/2021).

[124] S Hessler et al. "Hemocompatibility of EpoCore/EpoClad photoresists on COC substrate for optofluidic integrated Bragg sensors". In: *Sensors and Actuators B: Chemical* 239 (2017), pp. 916–922.

[125] E. Fiedler, N Haas, and T. Stieglitz. "Suitability of SU-8, EpoClad and EpoCore for flexible waveguides on implantable neural probes". In: *2014 36th Annual International Conference of the IEEE Engineering in Medicine and Biology Society*. IEEE. 2014, pp. 438–441.

[126] K Norrman, A Ghanbari-Siahkali, and N. Larsen. "6 Studies of spin-coated polymer films". In: *Annual Reports Section" C"(Physical Chemistry)* 101 (2005), pp. 174–201.

[127] G Yu and A. Heeger. "Charge separation and photovoltaic conversion in polymer composites with internal donor/acceptor heterojunctions". In: *Journal of Applied Physics* 78.7 (1995), pp. 4510–4515.

[128] J. Halls et al. "Efficient photodiodes from interpenetrating polymer networks". In: *Nature* 376.6540 (1995), pp. 498–500.

[129] D. Bornside, C. Macosko, and L. Scriven. "Spin coating: One-dimensional model". In: *Journal of Applied Physics* 66.11 (1989), pp. 5185–5193.

[130] W. W. Flack et al. "A mathematical model for spin coating of polymer resists". In: *Journal of Applied Physics* 56.4 (1984), pp. 1199–1206.

[131] M. Petty. *Molecular electronics: from principles to practice. 2007*. Chichester, John Wiley Sons Ltda, 2007.

[132] Y. Huang et al. "Soft lithography replication of polymeric microring optical resonators". In: *Optics Express* 11.20 (2003), pp. 2452–2458.

[133] J. Arrue et al. "Polymer-optical-fiber lasers and amplifiers doped with organic dyes". In: *Polymers* 3.3 (2011), pp. 1162–1180.

[134] S. Spelthann et al. "Towards Highly Efficient Polymer Fiber Laser Sources for Integrated Photonic Sensors". In: *Sensors* 20.15 (2020), p. 4086.

[135] K. Kuriki et al. "High-efficiency organic dye-doped polymer optical fiber lasers". In: *Applied Physics Letters* 77.3 (2000), pp. 331–333.

[136] *Optical Profilometry*. URL: https://www.nanoscience.com/techniques/optical-profilometry/# (visited on 03/01/2021).

[137] K. Shaklee and R. Leheny. "Direct determination of optical gain in semiconductor crystals". In: *Applied Physics Letters* 18.11 (1971), pp. 475–477.

[138] M. Čehovski et al. "Combined optical gain and degradation measurements in DCM2 doped Tris-(8-hydroxyquinoline) aluminum thin-films". In: *Organic Photonics VII*. Vol. 9895. International Society for Optics and Photonics. 2016, p. 989508.

[139] M. R. Kaysir et al. "Optical gain characterization of Perylene Red-doped PMMA for different pump configurations". In: *Applied optics* 55.1 (2016), pp. 178–183.

[140] L. Dal Negro et al. "Applicability conditions and experimental analysis of the variable stripe length method for gain measurements". In: *Optics communications* 229.1-6 (2004), pp. 337–348.

[141] M. D. McGehee et al. "Amplified spontaneous emission from photopumped films of a conjugated polymer". In: *Physical Review B* 58.11 (1998), p. 7035.

[142] I. D. Samuel, E. B. Namdas, and G. A. Turnbull. "How to recognize lasing". In: *Nature Photonics* 3.10 (2009), pp. 546–549.

[143] K. Kuriki et al. "Organic dye-doped polymer optical fiber laser". In: *Polymers for Advanced Technologies* 11.8-12 (2000), pp. 612–616.

[144] G. D. Peng et al. "Dye-doped step-index polymer optical fiber for broadband optical amplification". In: *Journal of lightwave technology* 14.10 (1996), pp. 2215–2223.

[145] M. Sheeba et al. "Multimode laser emission from dye doped polymer optical fiber". In: *Applied optics* 46.33 (2007), pp. 8089–8094.

[146] F. M. Zehentbauer et al. "Fluorescence spectroscopy of Rhodamine 6G: concentration and solvent effects". In: *Spectrochimica Acta Part A: Molecular and Biomolecular Spectroscopy* 121 (2014), pp. 147–151.

[147] H. Zhang et al. "Random lasing from Rhodamine 6G doped ethanediol solution based on the cicada wing nanocones". In: *Laser Physics* 26.6 (2016), p. 065004.

[148] A. Malinin et al. "Photonic crystal fibers for food quality analysis". In: *Biophotonics: Photonic Solutions for Better Health Care III*. Vol. 8427. International Society for Optics and Photonics. 2012, p. 842746.

www.ingramcontent.com/pod-product-compliance
Ingram Content Group UK Ltd.
Pitfield, Milton Keynes, MK11 3LW, UK
UKHW021653190726
13853UKWH00001B/230